Applications of Reference Materials in Analytical Chemistry

Applications of Reference Materials in Analytical Chemistry

Peter Roper, Shaun Burke, Richard Lawn, Vicki Barwick and Ron Walker

Laboratory of the Government Chemist, Teddington, UK

The work described in this book was supported under contract with the Department of Trade and Industry as part of the National Measurement System's Valid Analytical Measurement (VAM) Programme.

A catalogue record for this book is available from the British Library

ISBN 0-85404-448-5

Published for Laboratory of the Government Chemist
by the Royal Society of Chemistry,
Thomas Graham House, Science Park, Milton Road, Cambridge CB4 0WF, UK
Registered Charity Number 207890

Typeset by Paston PrePress Ltd, Beccles, Suffolk
Printed and bound by Bookcraft Ltd, UK

Preface

In almost every walk of life, reliable analytical measurement results are crucial to making correct decisions. International trade, environment protection, law enforcement, consumer safety and human health all rely heavily on the results produced by analytical chemists. In these and other areas, poor quality analytical data generates a cost penalty due not only to the need for repeat measurements, but also to the consequences of inappropriate actions based on false data. In the latter situation, the costs involved can be extremely high.

In order to achieve reliable measurement results, it is clearly necessary that the results of a particular analysis produced in a single laboratory can be shown to be comparable to the same analysis produced in different laboratories and locations. One effective means of achieving this situation is to ensure that all analytical data is traceable to reliable certified reference materials (CRMs). A laboratory can check the repeatability of its data by setting up internal quality control procedures to provide evidence of day-to-day consistency in its results, but a laboratory relying exclusively on this approach could conceivably be producing consistent, but biased results. The use of CRMs as measurement benchmarks can provide the essential reference or anchor points against which analysts can achieve comparability of their measurement results.

Even so, it is not uncommon for some laboratory managers to justify the high expenditure of regularly upgrading to the latest type of analytical equipment, and yet be overly conservative in the purchase of suitable CRMs to ensure that methods and equipment are correctly validated and calibrated. Indeed, several articles have recently appeared in the scientific press expressing concern that many analytical laboratories still do not use CRMs and that scientific journals covering analytical quality issues continue to accept papers that make no reference to their use in the work carried out. Given the increasingly higher profile that has been placed on the importance of analytical quality during the past decade this situation is disappointing.

One reason for the apparent reluctance to use CRMs on a regular basis could be that there is relatively little guidance available on their correct use. It is for this reason that this book has been written. It provides guidance on how CRMs can best be used to achieve analytical measurement results that are fit for purpose in a way that the reader should find straightforward and under-standable.

The main aim in writing this book is to explain how the results obtained from the use of CRMs can best be interpreted. The key applications of CRMs in analytical chemistry, including instrument calibration, method validation, checking laboratory performance, and assessing the accuracy of analytical data are described. Numerous worked examples are used to illustrate the principles involved and to lead the analyst through the most common calculations and statistical tests employed. Each chapter describes a particular use of CRMs and is written so that it may be read independent of other chapters. There is, therefore, some overlap in the content and topics discussed. However, it is hoped that the reader will find this approach beneficial.

The book has been produced as part of the UK Valid Analytical Measurement (VAM) programme. VAM is a programme of work funded by the Department of Trade and Industry as part of the UK's National Measurement System.

Ron Walker

Contents

CHAPTER 1

Introduction

This book has been produced as a deliverable under the UK's Valid Analytical Measurement (VAM) Programme. VAM is a programme of work sponsored by the Department of Trade and Industry as part of the UK's National Measurement System. The book provides guidance and information on the application of certified reference materials (CRMs) in the context of how they can best be used to achieve valid analytical measurements and thereby improve quality in the analytical laboratory.

The main applications of CRMs in analytical chemistry, such as instrument calibration, method validation, checking laboratory performance, internal quality control and uncertainty estimation are described. Worked examples are used to illustrate key issues and to lead the analyst through the most common calculations and statistical tests employed. The first two chapters provide general information on CRMs and how they are produced, whilst in Chapter 3 the statistics relating to the use of CRMs are explained in an easy to understand manner. Chapters 4–6 describe the main applications of CRMs, and are written so that they may be read independently of each other.

1.1 Certified Reference Materials and the VAM Principles

All chemical measurements, whether qualitative or quantitative, depend upon and are ultimately traceable to, a CRM or a standard material of some sort. Qualitative measurements of identity based, for example, on gas chromatographic retention times or spectroscopic properties, require a reliable, authentic, reference material to calibrate the particular instrument or test used. Quantitative determinations have the additional requirement that the instrument is calibrated with an accurately known amount of the reference material concerned.

The key role of CRMs in analytical chemistry is recognised in the VAM Principles,[1] a set of six statements, shown in the box, that set out the essential practical requirements for achieving valid analytical measurements.

The testing of methods and equipment referred to in VAM Principle 2 is most effectively accomplished by the use of appropriate CRMs. For example, the accuracy of the wavelength scale of UV-visible HPLC detectors can be verified

---

The VAM Principles

1. Analytical measurements should be made to satisfy an agreed requirement.
2. Analytical measurements should be made using methods and equipment that have been tested to ensure they are fit for their purpose.
3. Staff making analytical measurements should be both qualified and competent to undertake the task.
4. There should be a regular and independent assessment of the technical performance of a laboratory.
5. Analytical measurements made in one location should be consistent with those made elsewhere.
6. Organisations making analytical measurements should have well defined quality control and quality assurance procedures.

---

by the use of a CRM comprising a solution that has absorption peaks with well characterised reference values for the wavelengths of maximum absorbance. Likewise, the accuracy of an entire analytical method, such as the determination of the proximate constituents in foods, could be checked by the use of a CRM comprising a food-type matrix with well characterised reference values for constituents such as protein (nitrogen), moisture, fat, fibre and ash.

VAM Principle 5 emphasises the importance of comparability between analytical data produced in different laboratories and locations. One effective means of achieving this is to ensure that all analytical data are traceable to reliable CRMs. A laboratory can check the repeatability of its data by setting up internal quality control (IQC) procedures to provide evidence of day-to-day consistency in its results, but a laboratory relying exclusively on IQC procedures could conceivably be producing consistent, but biased results. The use of CRMs as measurement benchmarks can provide the essential reference or anchor points against which analysts can achieve comparability of their measurements. When several laboratories can achieve the same analytical result (within the uncertainty specified) for a given CRM, they will have demonstrated the comparability of their measurements.

1.2   Definitions and Hierarchy of Reference Materials

Although this book mainly covers the uses of CRMs, they are essentially a subset of materials that are generically known as reference materials. Within the scope of the generic term there are several classifications, and the following international definitions for some of these have been published by the International Organisation for Standardisation (ISO):[2]

- **Primary Standard:** Standard that is designated or widely acknowledged as having the highest metrological qualities and whose value is accepted without reference to other standards of the same quantity.

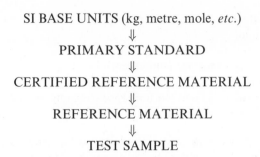

SI BASE UNITS (kg, metre, mole, *etc.*)
⇓
PRIMARY STANDARD
⇓
CERTIFIED REFERENCE MATERIAL
⇓
REFERENCE MATERIAL
⇓
TEST SAMPLE

Figure 1.1   *Hierarchy of reference materials*

- **Secondary Standard:** Standard whose value is assigned by comparison with a primary standard of the same quantity.
- **Certified Reference Material (CRM):** Reference material, accompanied by a certificate, one or more of whose property values are certified by a procedure which establishes its traceability to an accurate realisation of the unit in which the property values are expressed, and for which each certified value is accompanied by an uncertainty at a stated level of confidence.
- **Reference Material (RM):** Material or substance one or more of whose property values are sufficiently homogeneous and well established to be used for the calibration of an apparatus, the assessment of a measurement method, or for assigning values to materials.

These various types of reference materials can be conveniently arranged, as shown in Figure 1.1, according to their relative metrological positions (essentially, the uncertainty associated with their measurement value) between the SI base units and routine test sample measurements.

It should be noted that the position of a particular reference material in the hierarchy is not necessarily an indication of its suitability for a particular purpose. For example, in the determination of trace copper in a soil, a matrix CRM comprising a soil with certified levels of copper would be of greater value than a primary standard consisting of ultra-pure copper.

Primary standards represent the top-tier of chemical standards and, in principle, provide a means of establishing the traceability of analytical data to the SI measurement units, such as the kilogram, the metre and the mole. The concept of primary standards in analytical chemistry has been in existence for many years, and a detailed system for establishing the traceability of a range of pure chemicals and their solutions to a primary standard of ultra-pure silver was published in 1950.[3] Subsequently, the characteristics of primary standards were defined in more detail,[4] and now a number of pure chemicals are recognised as primary standards.

The essential characteristics of a primary standard are that it should be:

- readily available commercially

- of high chemical purity 100% ± 0.02%
- of high stability
- homogeneous
- non-hygroscopic and non-efflorescent
- readily soluble
- of high equivalent weight (to minimise weighing errors)
- able to undergo accurate stoichiometric reaction in titration

In addition, variations in isotopic abundance should not materially affect the molecular weight. A number of pure chemicals are recognised as primary standards including pure silver, zinc, bismuth, sodium carbonate, potassium dichromate, potassium iodate, iodine, sodium chloride and sulfamic acid. The system of primary standards and the traceability that ensues from it is essentially based on classical volumetric and gravimetric analysis. However, the system does extend to certain types of trace analysis, such as trace metal analysis. It should be noted that most CRMs are not primary standards although, where practical, the characterisation of the property values of candidate CRMs should be carried out using procedures that are traceable to primary standards.

1.3 Types of Reference Material

In terms of chemical composition, there are two main types of reference material:

- single substance reference materials (including primary standards)
- matrix reference materials

Single substance reference materials are pure chemicals (elements or compounds), or solutions of pure chemicals that have well-characterised reference values for such properties as chemical purity, concentration, melting point, enthalpy of fusion, viscosity, UV-visible absorbance, flash point, *etc.* An important use of reference materials of this type is in the calibration of analytical instrumentation and, as such, they feature in the great majority of analytical determinations. The choice of which reference material to use will depend on a variety of factors including availability, cost, suitability and the measurement uncertainty required for the measurements that are ultimately based on the use of that material.

Matrix reference materials are usually real-world materials containing the analytes of interest in their natural form and in their natural environment. Matrix reference materials should be chosen which have a matrix that closely resembles the matrix of the samples to be tested. In addition, they should ideally contain analytes with well-characterised reference values that are similar to the samples to be tested. The most important use of matrix reference materials is in the testing and validation of analytical methods. In contrast to single substance reference materials, which are primarily used in the measurement steps of an analytical process (*i.e.* for instrument calibration), matrix reference materials are introduced at the beginning of the analytical process. They are therefore used to assess the quality of the entire analytical process including sample extraction, clean up and concentration, as well as the final measurement step.

1.4 Uses of Reference Materials in Analytical Chemistry

Reference materials play an important role in analytical chemistry, their uses[5] including:

- checking instrument performance (wavelength, absorbance, melting point, *etc.*)
- calibrating instruments
- validating methods and estimating the uncertainty of analytical measurements
- checking laboratory and analyst performance
- internal quality control

In general, CRMs should be used within the framework of a comprehensive quality assurance system. It is not acceptable to use a CRM once only and then to assume that accurate results will be produced in the future. There are, however, cases for the legitimate use of CRMs on a non-routine basis. For example, when a new method has been developed and information is needed on whether reliable and accurate results are being produced. Equally, it may be appropriate to use suitable RMs for everyday applications, once an analytical system has been properly validated and calibrated by the use of a CRM and a correlation that is fit for purpose has been derived between the results obtained with the RM and the CRM.

1.5 Interpretation of Results Obtained with CRMs

Many analysts are unsure as to how best to interpret the results obtained from the use of a CRM. This includes how to compare the actual test results obtained when using the CRM with its certified reference values and their associated uncertainties.

The most common problems voiced by analysts include:

- How many replicate measurements are required for a proper comparison of the certified reference values and the test results obtained when analysing the CRM?
- Is it necessary for the mean of the test results found for a CRM to lie within the uncertainty range of the certified value, and are any differences found actually significant?
- If the mean of the test results found for a CRM lies outside the uncertainty range, what valid conclusions can be drawn and what action should be taken (if any)?

- Is a result outside the uncertainty range acceptable and if so what are the limits that do have to be respected? What are the conclusions and necessary actions in case of non-compliance with these limits?

The answers to these questions are covered in Chapters 4–6.

1.6   Availability of Reference Materials

The huge range of different sample matrices, requiring many different analytical determinations to be carried out, results in hundreds of thousands of measurements being made every year in a wide range of laboratories. It therefore follows that the number and range of reference materials that are required is equally large and diverse.

It is the analyst's responsibility to choose the best available material for the particular requirement. Information on the availability of reference materials and CRMs can be obtained from a number of sources including:

- The COMAR certified reference material database[6]
- Reference material producer catalogues
- Internet web sites of major reference material producers (*e.g.* LGC, NIST, BCR, *etc.*)

A list of the main CRM producers can be found in Annex A.

1.7   Quality of Certified Reference Materials

The absence of any internationally recognised classification or approval system for CRMs can make it difficult for the user to choose a reference material whose quality is appropriate for its intended use.

CRMs are supplied with certificates documenting their certified values and their associated uncertainties. The procedures by which the values were obtained and other information concerning homogeneity, stability and the correct use of the material may also be included. Some CRMs are also provided with detailed scientific reports that provide further detail. Most RMs and CRMs have well characterised property values, but some do not. It is important that a CRM has been produced and characterised in a technically valid manner. Users of CRMs need to be aware that not all materials have been validated to the same standard. Details of homogeneity trials, stability trials, method(s) used for characterisation and the uncertainties associated with the certified values are usually available from the producer and can, in part, be used to judge the pedigree and quality of the reference material. Also, guidance on CRM certificates,[7] as well as guidance on production procedures[8,9] is available from ISO, which may provide assistance in dealing with any enquiries to producers or suppliers.

Unfortunately, at present there is no independent means by which the users of CRMs can be assured of the adequacy of the processes used to certify the materials, or of the adequacy and integrity of the information contained in CRM certificates. There is great variability in certificates issued around the world and few means exist by which users can judge for themselves those materials which are adequate and fit for purpose. This is clearly an unsatisfactory situation and one that might seriously disadvantage those producers whose materials have been rigorously characterised. Some form of independent quality assessment of reference material producers, possibly including accreditation or registration, would both help users judge the quality of the reference materials offered by different producers and allow reference material producers to demonstrate the quality of their products. The recently published ISO Guide 34,[8] which provides guidance on the general quality requirements for the production of reference materials, is a step in this direction.

1.8 References

1. The VAM Principles, M. Sargent, *VAM Bulletin*, 1995, No. 13, pp. 4–5.
2. *Terms and Definitions Used in Connection with Reference Materials*, ISO Guide 30, 1992.
3. Standardisation of Volumetric Solutions, Analytical Chemists Committee of ICI, *Analyst*, 1950, **75**, 577.
4. Sodium Carbonate as a Primary Standard in Acid–Base Titrimetry, Analytical Standards Sub-Committee of the RSC, *Analyst*, 1965, **90**, 251.
5. *Uses of Certified Reference Materials*, ISO Guide 33, 1989 (under revision).
6. Information on the COMAR CRM database can be obtained at the COMAR web site (www.bam.de/comar), or the COMAR Central Secretariat, BAM, Rudower Chaussee 5, D-12489, Berlin.
7. *Contents of Certificates of Reference Materials*, ISO Guide 31, 1981.
8. *General Requirements for the Competence of Reference Material Producers*, ISO Guide 34, 2000.
9. *The Certification of Reference Materials – General and Statistical Principles*, ISO Guide 35, 1989.

CHAPTER 2

CRM Production

The process of preparing and characterising a CRM typically consists of a number of different steps including:

- assessment of priority needs and confirmation of demand
- literature search and project plan
- raw material selection and processing
- homogeneity testing
- stability testing
- characterisation
- preparation of the certification report and certificate

The sequence of main steps in CRM production is illustrated in Figure 2.1.

Each step in the production process will be discussed briefly in the following sections in order that the user can evaluate information on the suitability and quality of a given reference material more readily.

2.1 Assessment of Priority Needs and Confirmation of Demand

This is an essential first step in the production of a new CRM. Clearly, it is important to produce CRMs that meet the priority needs of the analytical community. Assessing this need is usually carried out by the use of surveys, questionnaires, customer liaison, and discussion at technical committees and conferences. For commercial production, the demand for a new CRM needs to be sufficient to ensure that the full production and marketing costs are recovered within the life span of the material. When government funding is available, such as that under the UK's Valid Analytical Measurement (VAM) Programme, the emphasis is usually placed on producing materials which are priority needs, but for which it would not be financially viable to produce commercially.

2.2 Literature Search and Project Plan

It is important to confirm that it is technically feasible to produce the required CRM and that any potential problems are identified, especially those requiring

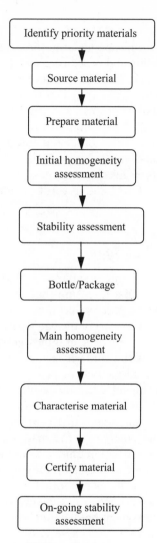

Figure 2.1 *Key to main steps in CRM production*

an element of research and development prior to starting production. Producing a new CRM can take from one to five years and proper planning is essential for success. A detailed project plan for production is important in setting out exactly what needs to be done, what resources are required and when.

2.3   Raw Material Selection and Processing

Identifying a source for a particular material with the desired property values is not always straightforward and may involve the producer in much initial research and consultation with outside bodies. Choosing the potential source of a raw material is often dependent on the quantity of final material required.

Commonly in the order of 100–200 kg of raw material is processed down to produce 1000–5000 units of finished material (unit sizes are typically 25–100 g for matrix materials). Sourcing and processing such large quantities of material requires specialised equipment, knowledge and expertise and is not within the normal scope of a typical chemical testing laboratory.

Most materials will need to be processed in some way to be suitable for use as a CRM. Processing may involve drying, crushing, milling, grinding, sieving, mixing, blending, riffling *etc.* Once processed, the material needs to be very thoroughly mixed and sub-divided for packaging into the individual units. Careful consideration needs be given to the type of containers used for the CRM such that the integrity of the CRM and its certified value(s) are maintained over a long time period.

2.4   Homogeneity Testing

Once packaged (*e.g.* bottled or ampouled) into individual units, it is vital to test a representative number of units to ensure that the material is homogeneous between units. That is, the property value(s) measured on one portion of the batch is comparable to any other part of the batch within acceptable uncertainty limits. For solid materials (*e.g.* powders, granules, *etc.*), it is usually necessary to establish homogeneity both within and between individual units. In general, it is important to use an analytical method or procedure that has good precision (repeatability). Producing highly accurate results is not essential, since the purpose of the exercise is to assess the difference (if any) in the property values between units. In practice, a material is often considered to be homogeneous if the difference between one portion of the material and another cannot be detected experimentally, *i.e.* the within unit variation and unit-to-unit variation are within the method variance. If this is not the case, then a contribution from the homogeneity assessment is incorporated into the total measurement uncertainty of the assigned (certified) values. However, even when no inhomogeneity is detected, it can be argued that it could be present but is masked by the method's repeatability. Therefore, in these cases a contribution to the total measurement uncertainty based on the method's variance should be included.

If a material is produced in several batches, it will be necessary to check the equivalence of the batches or to assign property values to each batch separately. Assessment of homogeneity should normally be performed after the material has been packaged into its final form unless stability studies indicate that storage should be maintained in the bulk form. In some cases, an intermediate homogeneity check may be necessary, *e.g.* prior to ampouling (see Figure 2.1).

2.5   Stability Testing

To be of value, a reference material must be stable for an acceptable time-span, under realistic conditions of storage, transport and use. Stability testing is an important part of the pre-production research and development process, to establish the storage conditions under which the candidate material is sufficiently stable to justify full-scale production.

For example, a material's stability to light, humidity and heat can be tested by storing replicate samples in different environments and analysing them after set periods of time. Accelerated stability testing studies on samples maintained at elevated temperatures are often employed to simulate the effects of long term storage. Generally, most stability studies involve storing samples at $-20\,°C$, at $+4\,°C$, at ambient temperature (approx. $+20\,°C$) and at $+40\,°C$, and then analysing them after periods of 0, 1, 3, 6 and 12 months. Occasionally, more sensitive materials, *e.g.* microbiological samples, may only be sufficiently stable to be used as CRMs if they are kept in storage at a temperature of $-70\,°C$ or below.

The nature of the container used to store the reference material can also play a significant role in the stability of a material. If not chosen correctly, the container material may interact with the material and adversely affect its long-term stability. The long-term stability of a CRM will also need to be monitored throughout its lifetime, *i.e.* its shelf life.

2.6 Characterisation

There are several technically valid approaches to characterising a CRM as described in ISO Guide 35.[1] These include using:

- gravimetric preparation data to estimate composition
- primary (definitive) methods
- independent measurement methods
- inter-laboratory studies

Depending on the type of CRM, its intended use, the competence of the laboratories involved and the quality of methods employed, one of the approaches may be chosen, as appropriate.

2.6.1 Characterisation Using Gravimetric Preparation Data

Certification of composition on the basis of gravimetric preparation data is preferred when it is feasible, but this is only the case exceptionally. Typical examples include the gravimetric preparation of gas mixtures or calibration solutions using certified pure or matrix CRMs containing accurately weighed amounts of an analyte(s) that has been thoroughly blended into the matrix. The traceability of a CRM prepared gravimetrically can be established by the traceability of the weights used to national standards of mass, the atomic/molecular masses of the components weighed out and their purity. Traceability to national standards of mass is relatively easy to achieve in theory, but in practice the mixtures prepared may not have the expected compositions because of unexpected instability, segregation, adsorption or other unidentified problems.

2.6.2   Characterisation Using a Primary (Definitive) Method

Characterisation of composition on the basis of analysis using a primary (definitive) method is metrologically sound, but relatively few methods exist. The use of primary methods is usually undertaken by a single laboratory employing a method that is based on first principles, *e.g.* gravimetry, volumetry, coulometry, isotope dilution mass spectrometry (IDMS) *etc*. Such methods have very high precision and very low or zero systematic error, thereby enabling the value of the parameter of interest to be measured within a narrow range of uncertainty. For example, IDMS has overcome many of the problems associated with the accurate determination of trace elements. Its capacity to compare number ratios of isotopic atoms of different masses whilst not requiring quantitative separation of the sample yields results which, in theory, are directly traceable to the mole. Spiking the sample with the isotopically labelled element or compound of interest, followed by the creation of conditions under which isotopic homogenisation can occur, enables the ratio of the analyte and spike to be determined mass spectrometrically under conditions which are essentially free from any matrix effects.

However, in practice most primary methods depend on some form of standard and they are applicable to relatively few analytical problems. They can also be extremely time-consuming and expensive but, when carried out correctly, can produce extremely high quality results.

2.6.3   Characterisation Using Independent Measurement Methods

Characterisation using independent measurement methods is most often undertaken by a single laboratory for the analysis of pure single substances such as pure pesticide CRM or gravimetrically prepared solutions requiring confirmatory analysis. Two or more well-validated analytical methods, each based on different fundamental principles, are used and if the results agree within their overall uncertainties, it provides a good indication that there is no bias in any of the methods used. The mean of the mean values can then be considered to be a reliable estimate of the true value. An example of this approach is the certification of a pure pesticide based on analysis by gas chromatography, high-pressure liquid chromatography and differential scanning calorimetry. In either case, the methods used should have small measurement uncertainties relative to the intended use of the CRM.

2.6.4   Characterisation by Inter-laboratory Studies

Characterisation collaborative trials (inter-laboratory consensus method) is the most commonly used approach to certifying reference materials, especially those based on natural matrices. When possible, the methods used should not only be different in operational details, but should also be based on different principles. For the highest quality work, the results of one or several preliminary inter-laboratory studies, in which a material similar to the proposed CRM is used, are compared in order to verify that the results of all the methods agree to within their measurement uncertainties. When they do not agree and there is evidence of systematic error, further work is undertaken to eliminate the causes of the error prior to carrying out the actual characterisation study. When the results of the different methods do agree to within their measurement uncertainties, there is only a very small chance of a significant common systematic error. Since the sources of error in one laboratory are usually unrelated to those in another laboratory, the residual systematic errors will usually cancel out and a reasonable estimate of the true result should be obtained.

Characterisation by inter-laboratory testing pre-supposes the existence of a number of equally capable laboratories, employing methods that have been independently validated. This approach implies that the differences between individual results are solely statistical in nature and can therefore be treated by purely statistical procedures. Although this approach to characterisation is often unavoidable, it frequently provides results that, strictly speaking, are only comparable between laboratories. This can lead to apparent authority being given to wrong values, especially if a purely statistical treatment is allowed to predominate over chemical wisdom and judgement.

A sub-set of characterisation by collaborative study is the method-specific approach, when the participating laboratories essentially all use the same method to establish the property values of interest. In this case, the property values obtained are method dependent and the reference material certificate should make this clear. There are many examples where this approach is the only valid one, especially in the regulatory field, *e.g.* leachable levels of toxic metals in paint, flash points of flammable solvents, *etc.* Many of the CRMs used in clinical chemistry are certified by the results of reference methods. The catalytic activity of an enzyme is evaluated by its ability to increase the rate of a particular chemical reaction under specified conditions of pH, temperature and concentration. The importance of using CRMs rigorously traceable to a reference method for the calibration of routine hospital instrumentation has only recently been recognised and is now laid down in European legislation.

2.7 Preparation of the Certification Report and Certificate

Once the CRM has been appropriately characterised and its certified values determined, a certificate is prepared. ISO Guide 31[2] provides guidance on the content of CRM certificates and reports. The CRM certificate should, as a minimum, contain information on:

- name and address of the producer
- description and name of the material
- certified property values and their uncertainties
- date of certification and shelf life
- intended use and any restrictions on use

- stability, transport and storage conditions
- instructions for correct use
- method of preparation
- statement of homogeneity and stability
- method of characterisation
- list of participating laboratories (when appropriate)

Some CRM producers also provide a more detailed report on the preparation, characterisation and certification of the material that contains additional information on:

- source of the material
- detailed preparation procedures
- techniques used for the homogeneity, stability and characterisation testing
- measurement results obtained by individual laboratories and the methods used

The traceability of CRMs can range from an unbroken chain of calibrations back to the relevant SI base units, to the use of results obtained from a well validated reference method. In each case, the producer will need to include in the CRM certificate a statement of traceability indicating the principles and procedures on which the property values (together with their uncertainty values) are based.

2.8   References

1. *The Certification of Reference Materials – General and Statistical Principles*, ISO Guide 35, 1989 (in revision).
2. *Contents of Certificates of Reference Materials*, ISO Guide 31, 1981 (in revision).

CHAPTER 3

Simple Statistics for Users of CRMs

The statistical methods described in this chapter are the tools that allow the CRM user to compare a measurement result(s) to a certified value on the CRM certificate. In order to do this effectively there are a number of rules that need to be kept in mind. More detailed information on statistics in general can be found in reference 1.

3.1 Rules for Collecting Data from CRMs

Rule 1. Independent replicate results are required.

Independent replication requires that all the stages of the analytical process are carried out separately in order to obtain individual results, *i.e.* a replicate result should not be influenced by previous common activities. For example, independent replication could mean that a completely new instrument calibration is required for each replicate determination in order to remove the effect of a common calibration.

Rule 2. Analytical data should not be rounded until after the statistical tests have been carried out.

Whenever repeated analytical measurements are made there is always some variation in the results caused by, for example, changes in environmental conditions or a slight variation in the way the sample is introduced into an instrument. If the data are rounded before carrying out any statistical tests, the natural variation within the data set can be masked (artificially increased or decreased). As a result, the statistical tests may lose power or not work at all, *e.g.*

7.172, 7.175, 7.165, 7.183, 7.18   Mean = 7.175   Standard deviation = 0.007
7.2, 7.2, 7.2, 7.2, 7.2   Mean = 7.2   Standard deviation = 0

Figure 3.1   *Blob plots of the raw data*

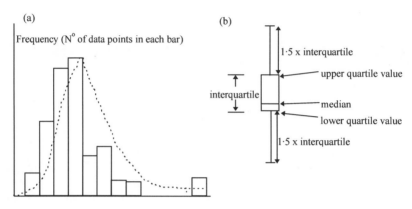

Figure 3.2   *Frequency histogram and Box and Whisker plot. The inter-quartile range is the range which contains the middle 50% of the data when it is sorted into ascending order (see also Section 3.4)*

Rule 3. Data should always be inspected before carrying out a statistical test.

Extreme results or incorrect assumptions about how the data are distributed can lead to false conclusions being reached. It is, therefore, always a good idea to plot the data as a first stage in the statistical analysis process.

For small amounts of data (< 15 replicate measurements), a blob (or dot) plot can be used to explore how the data are distributed, as shown in Figure 3.1. Blob plots are constructed by simply drawing a line, marking it off with a suitable scale and plotting the data along the axis.

For larger data sets frequency histograms (Figure 3.2a) and Box and Whisker plots (Figure 3.2b) may be better options to display the data distribution.

Rule 4. The analytical determinations should be made using a method that is under statistical control.

Wrong conclusions about the accuracy and precision of a method can be reached if the method being checked, validated or calibrated is not under statistical control. The replicate results should therefore be plotted against a time index (the order in which the data were obtained, *e.g.* hours, days, *etc.*). If any systematic trends are observed (Figure 3.3a–c) then the reasons for this must be investigated. Normal statistical methods assume a random distribution about the mean with time (Figure 3.3d).

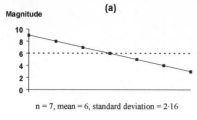

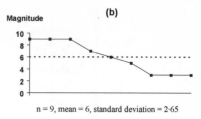

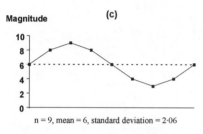

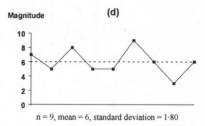

Figure 3.3 *Time indexed plots*

3.2 Concepts of Trueness, Precision and Accuracy

Trueness is the closeness of agreement between the average value obtained from a large set of test results and an accepted reference value. The measure of trueness is normally expressed in terms of bias.[2]

Precision is the closeness of agreement between repeated observations. It is usually expressed as a standard deviation.

Accuracy is the closeness of agreement between a test result and the accepted reference value. Note that the term, when applied to a set of test results, involves a combination of random components (*i.e.* precision) and a common systematic error or bias component.[2] It is a qualitative concept.

In Figure 3.4, the centre of the target is considered to be the true value. Good accuracy requires results to be closely grouped, *i.e.* precise, and their mean value to be close to the target value, *i.e.* true.

Looking at each target in Figure 3.4:

1. Lower left: results are scattered (poor precision), and their mean is some way off the centre of the target (poor trueness), overall poor accuracy.
2. Upper left: results are scattered (poor precision), but their mean is close to the centre of the target (good trueness), overall better accuracy than 1, but not as good as 4.
3. Lower right: results are closely grouped (good precision), but the mean is some way off the centre of the target (poor trueness), accuracy is better than 1, comparable with 2 but not as good as 4.
4. Upper right: results are closely grouped (good precision), and their mean is

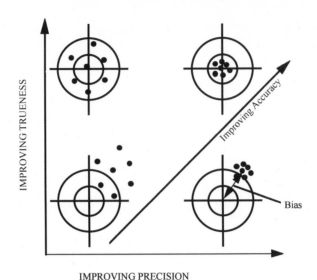

IMPROVING PRECISION

Figure 3.4 *The relationship between trueness, precision and accuracy*

close to the centre of the target (good trueness), overall the most accurate of the groups.

The precision measure of interest can be determined by replicate determinations of any suitable chemical standard (not necessarily a CRM). The only require-ments are that the standard used is stable and homogeneous and the user is clear what precision value is being estimated. For example, four different types of precision estimates are described in Table 3.1.

Precision is usually expressed as the sample standard deviation, s, calculated as follows:

$$s = \sqrt{\frac{\sum_{i=1}^{n}(x_i - \bar{x})^2}{n - 1}} \qquad \text{(Eq. 3.1)}$$

Table 3.1 *Different types of precision estimates*

	Description	Precision
1	Repeated injection of the same solution into the analytical instrument.	Injector precision
2	Repeat determination of a complete method by one analyst on the same day.	Repeatability (short-term precision)
3	Repeat determination of a complete method by several analysts over several days or weeks in the same laboratory.	Long term precision (intermediate precision)
4	Repeat determination of a complete method in different laboratories.	Reproducibility

where n is the number of replicates, x_i is the result for the ith replicate and $\bar{x}$ is the mean of the n replicates:

$$\bar{x} = \frac{\displaystyle\sum_{i=1}^{n} x_i}{n} \qquad \text{(Eq. 3.2)}$$

The summation symbol, sigma ($\sum_{i=1}^{n}$) simply means add up the values between $i = 1$ and n ($x_1 + x_2 + x_3 + x_4 + \ldots + x_n$).

Like precision, estimating method bias requires a number of replicate measurements to be made. Unlike precision, however, bias can only be assessed if the 'true value' of the measurand is known, at least to within a given degree of uncertainty.

Because of this requirement a CRM which has a known value for the measurand of interest is used to assess bias. [**Note**: When estimating bias the precision of the method can also be determined.]

3.3 Number of Required Replicates

To decide if the mean of a set of replicate results is equivalent to the certified value of a CRM, four factors must be taken into account:

- the number of replicate determinations made on the CRM using the method under investigation
- the precision of the method under investigation
- the uncertainty of the CRM's certified value
- the magnitude of the difference between the certified value of the CRM and the mean of the replicate results

Figure 3.5 shows how the estimate of the sample standard deviation (s) varies from that of the population standard deviation (σ) according to the number of replicates (n) from which the sample standard deviation is calculated. Essentially the more replicates determined the closer the ratio σ/s approaches unity, *i.e.* the sample standard deviation becomes a more perfect estimate of the population standard deviation. It can be seen from the graph that a small number of replicates will give a very poor estimate of the standard deviation. Normally, at least seven independent replicates are to be recommended. Likewise it can be seen that there is little to gain from carrying out more than 20 replicates.

To summarise, the sample mean and sample standard deviation calculated from replicate measurements are only estimates of the population (real) mean and standard deviation. How good an estimate they are depends on the precision of the method and on the number of replicate determinations made.

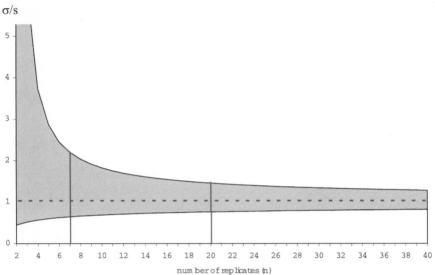

Figure 3.5   *Uncertainty associated with estimating a standard deviation*

3.4   Detection of Outliers

Occasionally, when we make repeated measurements, one or two results may appear anomalous, *i.e.* extreme. It is possible to check for the presence of these points of influence using one of the outlier tests described below, but the data should also be plotted (see Section 3.1, Rule 3).

Extreme values are defined as observations in a sample, so far separated in value from the remainder as to suggest that they may be from a different population, or the result of an error in measurement. Extreme values can also be sub-divided into *stragglers*, those extreme values detected between the 95% and 99% confidence level; and *outliers*, those extreme values detected at > 99% confidence level.

It is tempting to remove extreme values from a data set because it is believed their retention will incorrectly alter the calculated statistics, *e.g.* increase the estimate of precision, or possibly introduce a bias in the calculated mean. There is one golden rule, however, *no value should be removed from a data set on statistical grounds alone.* 'Statistical grounds' *include* outlier testing.

Outlier tests indicate, on the basis of some simple assumptions, where there is most likely to be a technical error. They do *not* indicate that a measurement value is incorrect. No matter how extreme a value is within a data set it could, nonetheless, be a correct piece of information. Only with experience, or the identification of a particular cause, can data be declared incorrect and removed from the statistical processing procedures.

Most outlier tests look at some measure of the relative distance of a suspect point from the mean value. This measure is then assessed to see if the extreme value could reasonably be expected to have arisen by chance. Most of the tests

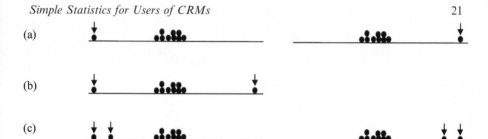

Figure 3.6   *Outliers and masking*

look for single extreme values (Figure 3.6a), but sometimes it is possible for several outliers to be present in the same data set. These can be identified in one of two ways:

- by iteratively applying the outlier test (not recommended when estimating the precision of an analytical method)
- by using tests which look for pairs of extreme values, *i.e.* outliers that are masking each other (*see* Figure 3.6b and 3.6c)

As a general rule of thumb, if more than 20% of the data is identified as outlying it is usually necessary to question the validity of the distribution of the data and/ or the quality of the data collected. For example, is the method under statistical control?

The appropriate outlier tests for the three situations illustrated in Figure 3.6 are:

(a) Grubbs 1
(b) Grubbs 2
(c) Grubbs 3

The test values for the three Grubbs' tests are calculated using Eq. 3.3 to Eq. 3.5, after arranging the data in ascending order.

$$G_1 = \frac{|\bar{x} - x_i|}{s} \qquad \text{(Eq. 3.3)}$$

$$G_2 = \frac{x_n - x_1}{s} \qquad \text{(Eq. 3.4)}$$

$$G_3 = 1 - \left(\frac{(n-3) \times s_{n-2}^2}{(n-1) \times s^2} \right) \qquad \text{(Eq. 3.5)}$$

where s is the standard deviation for the whole data set, x_i is the suspected single outlier, *i.e.* the value furthest away from the mean, $| \ |$ is the modulus – the value of a calculation ignoring the sign of the result, $\bar{x}$ is the mean, n is the number of data points, x_n and x_1 are the most extreme values, s_{n-2} is the standard

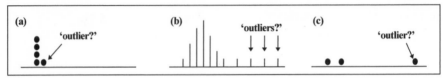

Figure 3.7   *Pitfalls of outlier testing*

deviation for the data set excluding the suspected pair of outlier values, *i.e.* the pair of values furthest away from the mean.

If the test values (G_1, G_2, G_3) are greater than the critical value for examples 3(a)–3(c) respectively, obtained from standard statistical tables (see Table A5, Annex B) then the extreme value(s) are unlikely to have occurred by chance at the stated confidence level.

Pitfalls of outlier tests

Figure 3.7 shows three situations where outlier tests can misleadingly identify an extreme value.

Figure 3.7a shows a situation common in chemical analysis. Since there is limited measurement precision (rounding errors) it is possible to end up comparing a result which, no matter how close it is to the other values, is an infinite number of standard deviations away from the mean of the remaining results. This value will therefore always be flagged as an outlier.

In Figure 3.7b there is a genuine long tail on the distribution which may cause successive outlying points to be identified. This type of distribution is surprisingly common in some types of chemical analysis.

If there are few measurement results (Figure 3.7c), an outlier can be identified by chance. In this situation it is possible that the identified point is closer to the 'true value' and that the other values are the outliers. This occurs more often than we would like to admit.

Worked example: Grubbs' tests
13 independent replicate determinations of wool content are carried out on a mixed polyester/wool fabric CRM. The results are arranged in ascending order.

x_1 x_n

47.876, 47.997, 48.065, 48.118, 48.151, 48.211, 48.251, 48.559, 48.634, 48.711, 49.005, 49.166, 49.484

$$n = 13, \text{mean} = 48.479, s = 0.498, s_{n-2}^2 = 0.123$$

$$G_1 = \frac{49.484 - 48.479}{0.498} = 2.02 \qquad G_2 = \frac{49.484 - 47.876}{0.498} = 3.23$$

$$G_3 = 1 - \left(\frac{10 \times 0.123}{12 \times 0.498^2}\right) = 0.587$$

The Grubbs' critical values for 13 determinations are $G_1 = 2.331$ and 2.607, $G_2 = 4.00$ and 4.24, $G_3 = 0.6705$ and 0.7667 for the 95% and 99% confidence levels respectively. Since, in all cases, the test values are less than their respective critical values it can be concluded that there are no outlying values.

3.5 Assessment of Precision

The precision of a method can be compared with a specification for the method using a Chi-squared test:

$$\chi_c^2 = \left(\frac{s}{\sigma}\right)^2 \qquad \text{(Eq. 3.6)}$$

where σ is the required precision value for the method expressed as a standard deviation, s is the determined method precision expressed as a standard deviation and χ_c^2 is the calculated Chi-squared test value.

The critical Chi-squared value is calculated as follows:

$$\chi_{\text{crit}}^2 = \frac{\chi_{(n-1);0.95}^2}{n-1} \qquad \text{(Eq. 3.7)}$$

where the value of $\chi_{(n-1);0.95}^2$ is found from standard statistical tables (see Table A4, Annex B, where df = degrees of freedom).

If $\chi_c^2 \leq \chi_{\text{crit}}^2$, there is no evidence that the precision of analytical method is inadequate.

If $\chi_c^2 > \chi_{\text{crit}}^2$, there is evidence that the precision of analytical method is inadequate.

Worked example
Consider the potentiometric titration of an $80 \, \text{mg} \, (100 \, \text{mL})^{-1}$ forensic ethanol in water CRM. The required precision expressed as a standard deviation is $0.1 \, \text{mg} \, (100 \, \text{mL})^{-1}$. The following results were obtained from 16 replicate determinations

79.99, 80.06, 80.04, 79.96, 80.03, 80.03, 80.01, 79.90, 80.03, 80.01, 79.89, 80.00, 79.89, 79.94, 79.98, 79.95 mg $(100 \, \text{mL})^{-1}$.

Mean $= 79.98124 \, \text{mg} \, (100 \, \text{mL})^{-1}$, Standard deviation $= 0.05450 \, \text{mg} \, (100 \, \text{mL})^{-1}$.

$$\chi_c^2 = \left(\frac{0.05450}{0.1}\right)^2 = 0.2970 \qquad \chi_{\text{crit}}^2 = \frac{24.999}{15} = 1.6666$$

Since $\chi_c^2 \leq \chi_{\text{crit}}^2$, there is no evidence that the precision of the potentiometric titration method for ethanol determination is inadequate.

If the precision of the method is found to be inadequate then the method should be looked at in detail to identify those areas where most variation is occurring. By controlling these areas first the greatest improvement in the method's precision can be achieved (see Concepts of Measurement Uncertainty, Section 3.10).

3.6 Assessment of Bias

Bias can be assessed by comparing the certified value, μ, with the mean result, $\bar{x}$, from replicate determination of a CRM, obtained using the method under investigation.

There is essentially no evidence for a bias if the following criteria are met:

$$-a_2 - 2\sigma - \leq \bar{x} - \mu \leq a_1 + 2\sigma \qquad \text{(Eq. 3.8)}$$

where σ is the precision of the method under investigation and a_1 and a_2 are adjustment values chosen by the investigator which take into account economic and/or technical considerations, such as a known and accepted bias in the method. The reason for the choice of values for a_1 and a_2 must always accompany any statement on bias.

Equation 3.8 assumes that the uncertainty in the certified value is insignificant compared with the method precision. If this is not the case then the bias assessment criteria become:

$$-a_2 - 2\sqrt{u_{\text{CRM}}^2 + \sigma^2} - \leq \bar{x} - \mu \leq a_1 + 2\sqrt{u_{\text{CRM}}^2 + \sigma^2} \qquad \text{(Eq. 3.9)}$$

where u_{CRM} is the standard uncertainty for the CRM.

It can be seen from Eq. 3.9 that bias detection is limited to a minimum of twice the standard uncertainty of the certified value of the CRM. It is therefore incumbent on the user of the CRM to select a material that is capable of detecting a possible significant bias.

Note 1: The words 'significant' and 'significance' have a specific meaning in statistics. A significant difference indicates a difference that is unlikely to have occurred by chance at a given level of confidence. A significance test highlights differences unlikely to occur due to purely random variation.

Note 2: Significance is a function of sample size. Comparing very large samples will nearly always lead to a statistically significant difference. However, a statistically significant result is not necessarily of any practical importance.

Worked example

The long-term precision of a method used to determine lead in water, estimated using a CRM check sample, was 1.17 mg kg^{-1}. The mean of the replicate measurement results was 57.6 mg kg^{-1}. The certified value for the CRM is 60.1 mg kg^{-1} with an expanded uncertainty of 2.5 mg kg^{-1}. A coverage factor of $k = 2$ has been used to calculate the expanded uncertainty. Is there any evidence of a bias in the determination?

$$\bar{x} - \mu = 57.6 - 60.1 = -2.5 \text{ and } 2\sqrt{u_{CRM}^2 + \sigma^2} = 2 \times \sqrt{\left(\frac{2.5}{2}\right)^2 + 1.17^2} = 3.42$$

Using (Eq. 3.9) indicates that there is no evidence of a significant bias in the method ($-3.42 \leq -2.5 \leq +3.42$).

Note:

1. The expanded uncertainty quoted on the certificate needs to be converted into a standard uncertainty (standard deviation). This is achieved by dividing the expanded uncertainty by the coverage factor, *i.e.* $\frac{2.5}{2}$ (see Section 3.7).
2. It is assumed that enough replicates have been determined to obtain a good estimate of the method precision, σ (see Number of Required Replicates, Section 3.3).

If a discrepancy is found between the user's measurement and the certified value the reason should to be investigated.

3.7 Converting Expanded Uncertainties into Standard Uncertainties

Uncertainties on CRM certificates are usually expressed as expanded uncertainties, U, and represent the limits within which the true value of the measurand is thought to lie with a stated degree of confidence. It is important to know how the expanded uncertainty is calculated so that it can be converted back to a standard uncertainty, u, for use in statistical tests.

Expanded uncertainties, expressed as 95% or 99% confidence limits, are converted into standard uncertainties as follows:

$$u = U_{95\%}/2 \qquad \text{(Eq. 3.10)}$$

$$u = U_{99\%}/3 \qquad \text{(Eq. 3.11)}$$

Expanded uncertainties derived from the use of a coverage factor, k, are converted into a standard uncertainty as follows:

$$u = U/k \qquad \text{(Eq. 3.12)}$$

3.8 Using CRMs to Detect Other Forms of Bias

In Section 3.6 it was shown how a significance test can be used to detect bias in a single method using a CRM. Significance tests can also be used to see if there is a difference between:

- the mean of replicate results and a regulatory limit
- two individual analysts

- two different methods
- individual laboratories within a single organisation
- two laboratories in different organisations
- the spread/dispersion of two sets of replicate data

Statistical terminology

There is much jargon surrounding significance testing and the CRM user needs to be familiar with at least some of it. For this reason, definitions of most of the terms generally encountered are summarised in Table 3.2.

3.8.1   Some Assumptions Behind Significance Testing

Like most statistical tests, it is assumed that the sample taken correctly represents the population and that the population follows a normal distribution. Although these assumptions are never complied with exactly, in the majority of situations where laboratory data are being used they are not grossly violated.

Table 3.2   *t-Test terminology*

Term	Definition
Alternative hypothesis (H_1)	A statement describing the alternative to the null hypothesis, *i.e.* there is a difference between the population means μ_1 and μ_2 (see two-tailed), or μ_1 is $> \mu_2$ (see one-tailed).
Critical value (t_{crit} or F_{crit})	The value obtained from statistical tables or statistical packages, at a given confidence level, with which the result of applying a significance test is compared.
Null hypothesis (H_0)	A statement describing what is being tested, *i.e.* there is no difference between the two population means ($\mu_1 = \mu_2$).
One-tailed	A one-tailed test is carried out when the analyst is only interested in the answer if the result is different in one direction. For example, (1) the new production method results in a higher yield; or (2) the amount of waste product is reduced (*i.e.* a limit value $<$ or $>$ is used in the alternative hypothesis). In these cases the calculation to determine the *t*-value is the same as that for the two-tailed *t*-test, but the critical value is different.
Population	A large group of items or measurements under investigation, *e.g.* a single batch of a CRM comprising 2500 units.
Sample	A group of items or measurements taken from the population, *e.g.* 25 units of a CRM taken from a batch containing 2500 units.
Two-tailed	A two-tailed *t*-test is carried out when the analyst is interested in any change regardless of direction. For example, is method (A) different to method (B), (*i.e.* $\neq$ is used in the alternative hypothesis).

Table 3.3 *t-Test formulae*

t-Test to use when comparing	*Equation*	
The long term average or regulatory limit ($\bar{x}_0$) with a sample mean	$$t = \frac{\bar{x} - \bar{x}_0}{s/\sqrt{n}}$$ The degrees of freedom v is given by: $$v = n_1 - 1$$	(Eq. 3.13)
The difference between pairs of analytical results (*e.g.* results obtained from two analytical methods)	For a two-tailed test, $t = \dfrac{|\bar{d}| \times \sqrt{n}}{s_d}$ The degrees of freedom v is given by: $$v = n_1 - 1$$	(Eq. 3.14)
Difference between independent sample means (assuming equal variance)	$$t = \frac{\bar{x}_1 - \bar{x}_2}{s_c\sqrt{\left(\dfrac{1}{n_1} + \dfrac{1}{n_1}\right)}}$$ The degrees of freedom v is given by: $$v = n_1 + n_2 - 2$$ and: $$s_c = \sqrt{\frac{s_1^2(n_1 - 1) + s_2^2(n_2 - 1)}{(n_1 + n_2 - 2)}}$$	(Eq. 3.15) (Eq. 3.16)
Difference between independent sample means (assuming unequal variance)	$$t = \frac{\bar{x}_1 - \bar{x}_2}{\sqrt{\dfrac{s_1^2}{n_1} + \dfrac{s_2^2}{n_2}}}$$ The degrees of freedom v is given by: $$\frac{1}{v} = \frac{s_1^4}{k^2 n_1^2(n_1 - 1)} + \frac{s_2^4}{k^2 n_2^2(n_2 - 1)}$$ where: $$k = \frac{s_1^2}{n_1} + \frac{s_2^2}{n_2}$$	(Eq. 3.17) (Eq. 3.18)

3.8.2 *t*-Tests

When comparing the mean result of a number of replicate determinations with another mean or regulatory limit, *t*-tests are used. There are a number of these, the formulae for each are given in Table 3.3. To carry out a *t*-test, the following steps are required:

1. State exactly what question is being asked. This is done in the form of two hypotheses: (i) the results being compared could really be the same (null

hypothesis), (ii) the results could really be different (alternative hypothesis). In statistical terminology this is written as:-

The null hypothesis (H_0): $\mu_1 = \mu_2$.

The alternative hypothesis (H_1): $\mu_1 \neq \mu_2$ (two-tailed) or $\mu_1 < \mu_2$ (one-tailed), or $\mu_1 > \mu_2$ (one-tailed).

2. Decide which of the *t*-tests listed in Table 3.3 is appropriate and calculate the *t*-value or statistic.
3. Look up the critical *t*-value (Table A1, Annex B). To do this three pieces of information are needed:
    (i)   is the direction of any difference between the two values being compared critical (one-sided), or is it only important to check for a statistically significant difference (two-sided)?
    (ii) the degrees of freedom for the appropriate *t*-test, calculated as per Table 3.3.
    (iii) how certain does the user need to be about the conclusions of the test? It is normal practice in chemistry to select the 95% confidence level.

    Note: In some situations this is an unacceptable level of error, *e.g.* in medical research. If necessary the 99% or even the 99.9% confidence level can be chosen.

    Using this information the critical *t*-value for the test can be found from Table A1, Annex B. For example, the critical *t*-value for a two-tailed test carried out at the 95% confidence interval with 14 degrees of freedom (15 replicate determinations) is 2.145.

In Table 3.3, $\bar{x}$ is the sample mean, μ is the population mean, *s* is the standard deviation for the sample, *n* is the number items in the sample, $|\bar{d}|$ is the absolute difference between paired means, $\bar{d}$ is the difference between paired means, s_d is the sample standard deviation for the pairs, $\bar{x}_1$ and $\bar{x}_2$ are two independent sample means, n_1 and n_2 are the number of items making up each sample, s_c is the combined standard deviation and s_1 and s_2 are the sample standard deviations for the two independent sample means.

Note: For a one tailed *t*-test the sign of the calculated *t* value is important, because it shows the direction of the difference. For a two-tailed *t*-test the sign of the calculated *t* value is ignored.

3.8.3   The *F*-Test

A different significance test, the *F*-test, compares the spread of results in two data sets. This is done in order to determine if they could reasonably be considered to come from the same parent distribution. The test can therefore be used to answer questions like:

- are two methods equally precise?
- can two estimates of spread (dispersion) be combined?

The measure of spread used in the *F*-test is *variance*, that is the square of the standard deviation. The ratio of the two variances is used to obtain the test value:

$$F = \frac{s_a^2}{s_b^2} \qquad \text{(Eq. 3.19)}$$

This value is then compared with a critical value from standard statistical tables to find out whether the difference in spread has occurred by chance. The F_{crit} value is found from tables using $(n_1 - 1)$ and $(n_2 - 1)$ degrees of freedom.

Note:

(1) If we are interested in the question 'is there a significant difference in the spread of the two sets of replicates?' (*i.e.* a two-tailed *F*-test), then it is normal practice to arrange s_1 and s_2 so that *F* is >1 and look up the 97.5% confidence level *F* critical value (Table A2, Annex B). Note that the confidence level of the test is still 95%.

(2) If we are interested in the question 'is the new method the same as or significantly more precise than the old method?' (*i.e.* a one-tailed *F*-test), then the test is as follows:

$$F = \frac{s_{new\ method}^2}{s_{old\ method}^2} \qquad \text{(Eq. 3.20)}$$

In this case the two precision estimates are not rearranged and the 95% confidence level *F* critical value is used (Table A3, Annex B).

In either case, the standard deviations are considered to be significantly different if $F > F_{crit}$.

Worked example: Testing against a set limit

An analyst is asked to determine if the results from a sample exceed the regulatory limit of 22.7 mg dm^{-3}. The mean of 10 replicate results is 23.5 mg dm^{-3}, with a standard deviation of 0.9 mg dm^{-3} (it is assumed that the analyst is competent and reports unbiased results).

To answer this question we use a *t*-test to compare the mean value with a set limit (see Eq. 3.13).

The null hypothesis (H$_0$): $\bar{x} =$ set limit ($\bar{x}_0$).

The alternative hypothesis (H$_1$): $\bar{x} >$ set limit ($\bar{x}_0$).

$$t\text{-value} = \frac{23.5 - 22.7}{0.9/\sqrt{10}} = +2.81$$

Because we are only interested in whether or not the sample is above the regulatory limit we carry out a one-tailed *t*-test.

$t_{crit} = 1.83$ one-tailed at the 95% confidence level for 9 degrees of freedom.

Since $t_{calculated} > t_{crit}$ we can reject the null hypothesis and conclude that the sample is significantly above the regulatory limit.

Worked example: Comparing independent methods

Two methods for determining the concentration of selenium have been compared using an environmental CRM. The results from each method are as follows:

						$\bar{x}$	s
Method 1	4.2	4.5	6.8	7.2	4.3	5.40	1.471
Method 2	9.2	4.0	1.9	5.2	3.5	4.76	2.750

Using the *t*-test for independent sample means we define the null hypothesis (H_0): $\mu_1 = \mu_2$, *i.e.* there is no difference between the means of the two methods (the alternative hypothesis is (H_1): $\mu_1 \neq \mu_2$). If the two methods have standard deviations which are not significantly different (checked using the *F*-Test – see Eq. 3.19), then we can combine (or pool) the standard deviations (s_c) and use the *t*-test for equal variance as shown below:

$$F = \frac{2.75^2}{1.471^2} = 3.49 \quad F_{crit} = 9.605 \ (5-1) \text{ and } (5-1) \text{ degrees of freedom.}$$

Since $F_{crit} > F_{calculated}$, we can conclude that the spread of results in the two data sets are not significantly different and it is therefore reasonable to combine the two standard deviations (see Eq. 3.16):

$$s_c = \sqrt{\left(\frac{1.471^2 \times (5-1) + 2.750^2 \times (5-1)}{(5+5-2)}\right)} = 2.205$$

Evaluating the test statistic *t* (see Eq. 3.15):

$$t \approx \frac{(5.40 - 4.76)}{\sqrt{\left(\frac{1}{5} + \frac{1}{5}\right)\left(\frac{1.471^2 \times (5-1) + 2.750^2 \times (5-1)}{(5+5-2)}\right)}} \Rightarrow$$

$$t \approx \frac{0.64}{\sqrt{0.4 \times \dfrac{8.6600 + 30.252}{8}}} \approx \frac{0.64}{\sqrt{1.946}} \approx 0.459$$

The 95% two-sided critical value is 2.306 for $v = 8(n_1 + n_2 - 2)$ degrees of freedom. This exceeds the calculated value of 0.459, so the null hypothesis (H_0) cannot be rejected and we conclude that there is no significant difference between the results given by the two methods, for the standard being tested.

Worked example: Paired difference

Two methods are available for determining the concentration of a vitamin in foodstuffs. In order to compare the methods, several CRMs with different matrices are prepared for analysis using the same extraction technique. Each sample preparation is then divided into two aliquots and readings are obtained on them by the two methods, ideally commencing at the same time to reduce the possible effects of sample deterioration. The results are as follows:

Sample i.d.	1	2	3	4	5	6	7	8
Method A	2.52	3.13	4.33	2.25	2.79	3.04	2.19	2.16
Method B	3.17	5.00	4.03	2.38	3.68	2.94	2.83	2.18
Difference (*d*)	−0.65	−1.87	0.30	−0.13	−0.89	0.10	−0.64	−0.02

The null hypothesis is (H_0): $\mu_d = 0$ against the alternative (H_1): $\mu_d \neq 0$, where μ_d is the population paired difference.

The test is a two tailed test as we are interested in both $0 < \bar{d}$ and $\bar{d} < 0$.

The mean, $\bar{d}$, $= 0.475$ and the sample standard deviation for the paired differences, $s_d = 0.700$.

The test statistic is $t = \dfrac{|\bar{d}| \times \sqrt{n}}{s_d}$ Evaluating gives $t = \dfrac{|0.475| \times \sqrt{8}}{0.700} = 1.918$

The tabulated value of t_{crit} (with $v = 7$, at the 95% two-sided confidence limit) is 2.365. Since the calculated value is less than the critical value, (H_0) cannot be rejected and it follows that there is no statistically significant difference between the two methods.

3.9 Using Statistical Software (What Is a *p*-Value?)

If statistical software packages (or spreadsheet built-in functions) are used to calculate the test value for the significance tests, a *p*-value is often reported. The *p*-value represents an inverse index of the reliability of the statistic (*i.e.* the probability of an error in accepting the observed result as valid). Thus, if two means are compared to see if they are different, a *p*-value can be used to calculate at what confidence level the difference is just significant:

$$\text{Confidence level } \% = 100 \times (1 - p) \qquad \text{(Eq. 3.21)}$$

It is usual when analysing chemical data (but somewhat arbitrary) to say that *p*-levels ≤ 0.05 are statistically significant.

3.10 Concepts of Measurement Uncertainty

Measurement uncertainty is defined as 'a parameter associated with the result of a measurement that characterises the dispersion of values that could reasonably be attributed to the measurand'.[3]

Put more simply, it is the value after the $\pm$, *e.g.* 25 ± 0.9 mg kg^{-1} potassium means the 'true' amount of potassium in the sample could reasonably be expected to lie between 24.1 and 25.9 mg kg^{-1}.

A large number of factors can affect the result of an analytical determination, all of which are potential sources of uncertainty. Each of the factors that may affect the result need to be considered in turn, be they due to possible random variations such as fluctuations in an oven temperature, or systematic variations such as a balance always reading 0.1 g too high.

Note: Gross-errors such as wrongly transcribed results or using the wrong reagent concentration are not included. These types of errors mean that the analytical process is not under statistical control.

ISO have produced an internationally accepted guide to estimating measurement uncertainty.[3] The guide recommends that the individual uncertainty estimates, expressed as standard uncertainties (u) or standard deviations (s), be combined in quadrature to yield the combined standard uncertainty, u_c.

That is to say, if the individual components of a method $(1 - n)$ are investigated to obtain separate estimates, the total uncertainty will be given by the following expression:

$$u_c = s_{total} = \sqrt{s_1^2 + s_2^2 + s_3^2 + \ldots + s_n^2} \qquad \text{(Eq. 3.22)}$$

One of the consequences of this approach of combining uncertainties is that the smaller estimates, *i.e.* those that are less than one third of the largest estimate, make a negligible contribution to the total uncertainty.

In most cases the reported uncertainty (the one found on most CRM certificates) is expressed as an expanded uncertainty, U. This is calculated by multiplying the combined standard uncertainty, u_c by a coverage factor, k:

$$U = k \times u_c \qquad \text{(Eq. 3.23)}$$

For most purposes, $k = 2$ is considered appropriate and gives a range approximating to the 95% confidence interval. For more demanding applications $k = 3$ is recommended.

Sometimes expanded uncertainties are reported directly as confidence intervals (either at 95% or 99% confidence) on CRM certificates.

3.11   Principles of Least Squares Linear Regression

Least squares linear regression is the statistical method used to summarise the degree of association between two variables, for example, measurement of the UV absorbances of a set of solutions using a spectrophotometer. The method works by finding the best curve through the data which minimises the sum of squares of the residuals (a residual being the difference between a datum point and the corresponding value on the curve [line]).

The normal equations

The method of least squares ensures that the sum of squares of the vertical deviations about the line, $\sum_{i=1}^{n}(y_i - \hat{y}_i)^2$, is minimised, where $\hat{y}_i$ is the estimated y value for the ith point.

To calculate the intercept (c) and slope (m) for the linear model the two normal equations are solved simultaneously:

$$\sum_{i=1}^{n} y_i = nc + m \sum_{i=1}^{n} x_i \qquad \text{(Eq. 3.24)}$$

$$\sum_{i=1}^{n} x_i y_i = c \sum_{i=1}^{n} x_i + m \sum_{i=1}^{n} x_i^2 \qquad \text{(Eq. 3.25)}$$

where x_i and y_i are the individual data values and n is the number of points in the regression.

Least squares linear regression is normally most easily carried out using standard statistical software. Regression analyses using these packages will provide a number of regression statistics such as slope, intercept and the correlation coefficient. The following sections explain these statistics in more detail.

Note: There are other regression methods such as ranked regression, multiple linear regression, non-linear regression, principal component regression, partial least squares regression, *etc.*, which are useful for analysing instrument/chemically derived data, but are beyond the scope of this introductory text.

3.11.1 Least Squares Linear Regression Statistics

Correlation Coefficient

The correlation coefficient (r) is a measure of the degree of association or linear relationship between two variables. The correlation coefficient ranges from -1, a perfect negative relationship, through zero (no relationship), to $+1$, a perfect positive relationship; see Figure 3.8a–c.

The value of r is, however, open to mis-interpretation. Figures 3.9a and 3.9b, for example, show instances where the r values alone would give the wrong impression of the underlying relationship. It is therefore essential to plot the data in order to check that linear least squares statistics are appropriate.

The statistical significance of the correlation coefficient is dependent on the number of data points in the calibration. To test whether a particular r value indicates a statistically significant relationship we can use the Pearson's correlation coefficient test (Table 3.4, Figure 3.10). For example, if we only have four points (degrees of freedom, $v = n-2 = 2$), an r value of -0.94 will not be significant at the 95% confidence level. However, if there are more than 60 data

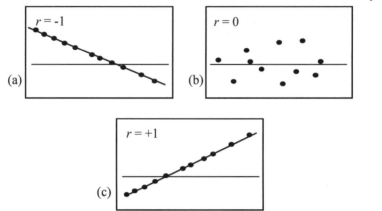

Figure 3.8   *The correlation coefficient (r)*

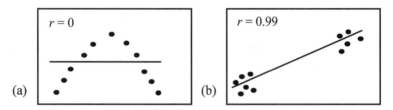

Figure 3.9   *Misinterpretation of r*

points ($v > 58$), an r value of just 0.26 ($r^2 = 0.0676$) would indicate a significant, but not very strong, positive linear relationship. Thus a linear relationship can be statistically significant, but of no practical value.

Also, it should be noted that the test used here simply shows whether two sets are linearly related; it does not 'prove' linearity or adequacy of fit.

Evaluating the correlation coefficient (r) involves using a complex formula. This calculation is almost invariably performed using standard statistical software packages.

Slope and Intercept

In linear regression the relationship between the x and y data is assumed to be represented by a straight line (see Eq. 3.26 and Figure 3.11),

$$y = mx + c \qquad\qquad \text{(Eq. 3.26)}$$

where y is the estimated response/dependent variable, m is the slope (gradient) of the regression line and c is the intercept (y value when $x = 0$).

The straight-line model is only appropriate if the data approximately fit the assumption of linearity. This can be tested for by plotting the data and looking

Table 3.4 *Pearson's correlation coefficient test*

Degrees of freedom (n−2)	Confidence level	
	95% (α = 0.05)	99% (α = 0.01)
2	0.950	0.990
3	0.878	0.959
4	0.811	0.917
5	0.754	0.875
6	0.707	0.834
7	0.666	0.798
8	0.632	0.765
9	0.602	0.735
10	0.576	0.708
11	0.553	0.684
12	0.532	0.661
13	0.514	0.641
14	0.497	0.623
15	0.482	0.606
20	0.423	0.537
30	0.349	0.449
40	0.304	0.393
60	0.250	0.325

Significant correlation when $|r| \geq$ table value

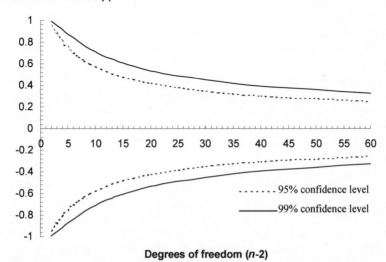

Figure 3.10 *Illustration of number of data points at which r becomes significant*

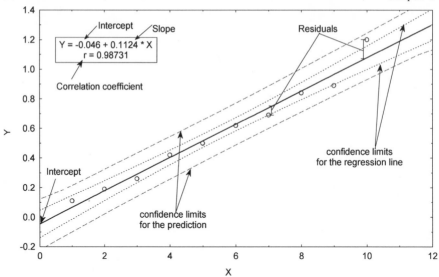

Figure 3.11   *A typical calibration plot*

for curvature, or by plotting the residuals against the predicted *y* values or *x* values (see below).

Residuals and Residual Standard Deviation

A residual value is calculated by taking the difference between the predicted value and the actual value (see Figure 3.11). When the residuals are plotted against the observed response data values the plot becomes a powerful diagnostic tool, enabling patterns and curvature in the data to be recognised (Figure 3.12). It can also be used to highlight points of influence (see bias, leverage and outliers below).

The residual standard deviation (rsd), also known as residual standard error (rse) or standard deviation of the line (sdl), is a statistical measure of the average residual, *i.e.* an estimate of the average error (or deviation) about the regression line. The rsd is used to calculate many useful regression statistics including confidence intervals and outlier test values.

$$\text{rsd} = s_y \sqrt{\frac{(n-1)}{(n-2)}(1 - r^2)} \qquad \text{(Eq. 3.27)}$$

where s_y is the standard deviation of the *y* values in the calibration, *n* is the number of data pairs and *r* is the least squares regression correlation coefficient.

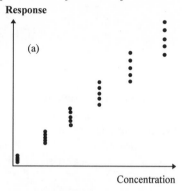

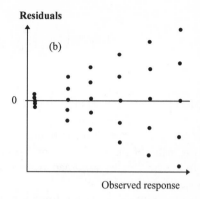

Figure 3.12 *Residuals plot*

3.11.2 Confidence Intervals

Like most statistics, the slope (m) and intercept (c) are estimates based on a finite sample, so there is some uncertainty[†] in the values. This uncertainty is quantified in a number of ways, *e.g.* standard deviations of the mean (standard errors), *t*-values, *p*-values and confidence limits. Examples of these statistics are given below for output generated using **MS Office EXCEL 97** (Table 3.5).

The confidence interval for the regression line can be plotted for all points along the *x*-axis and is dumb-bell in shape (Figure 3.11). In practice this means that the model is more certain in the middle than at the extremes, which in turn has important consequences for extrapolating relationships.

Table 3.5 *Regression statistics for the data used to generate the calibration graph (Figure 3.11)*

	Coefficients	Standard error	t-value	p-value	Lower 95%	Upper 95%
Intercept	−0.046	0.0396	−1.160	0.279	−0.1374	0.0454
Slope	0.112	0.0064	17.584	1.1E−07	0.0976	0.1271

Prediction Interval

When regression is used to construct a calibration model, the calibration graph is used in reverse, *i.e.* the *x* value is predicted from the instrument response (*y*-value). This prediction has an associated uncertainty (expressed as a confidence interval [CI]):

$$x_{predicted} = \left(\frac{(\bar{y}_0 - c)}{m} \right) \qquad \text{(Eq. 3.28)}$$

[†]Strictly, the uncertainty which arises from random variability between sets of data. There may be other uncertainties, such as measurement bias.

$$\text{CI is}\quad x_{\text{predicted}} \pm \left(\frac{t(\text{rsd})}{m}\right)\sqrt{\frac{1}{N} + \frac{1}{n} + \frac{(\bar{y}_0 - \bar{y})^2}{m^2(n-1)s_{(x)}^2}} \qquad \text{(Eq. 3.29)}$$

where:

c is the intercept and m is the slope obtained from the regression equation.

$\bar{y}_0$ is the mean value of the response (*e.g.* instrument readings) for N replicates (replicates are repeat measurements made at the same level).

$\bar{y}$ is the mean of the y data for the n points in the calibration.

t is the critical value obtained from t-tables for $n-2$ degrees of freedom at a stated confidence level.

$s_{(x)}$ is the standard deviation for the x data for the n points in the calibration.

rsd is the residual standard deviation for the calibration.

If a reduction in the size of the confidence interval of the prediction is required, there are several things that can be done:

1. Ensure that the unknown determinations of interest are close to the centre of the calibration, *i.e.* close to the values $\bar{x}$, $\bar{y}$ (the centroid point). This suggests that if we want a small confidence interval at low values of x, then the CRMs used in the calibration should be concentrated around this region. For example, a typical pattern of standard concentrations might be 0.05, 0.1, 0.2, 0.4, 0.8, 1.6 (*i.e.* only one or two standards are used at higher concentrations). While this will lead to a smaller confidence interval at lower concentrations, the calibration model will be prone to leverage errors (see below).
2. Increase the number of points in the calibration (n). There is, however, little improvement when increasing the number of different calibration points above 10 unless standard preparation and analysis is rapid and cheap.
3. Increase the number of replicate determinations for estimating the unknown (N). Once again there is a law of diminishing returns, so the number of replicates should typically be in the range 2–5.
4. The range of the calibration can be extended, providing the calibration is still linear.

3.11.3 Extrapolation and Interpolation

We have already mentioned that the regression line is subject to some uncertainty and that this uncertainty becomes greater at the extremes of the line. If we therefore try to extrapolate much beyond the point where we have real data ($\pm 10\%$), there may be relatively large errors associated with the predicted value. Conversely interpolation near the middle of the calibration will minimise the prediction uncertainty. It follows, therefore, that when constructing a calibration graph the standards should cover a larger range of concentra-

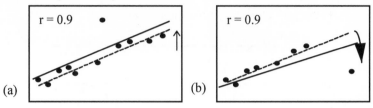

Figure 3.13 *Bias and leverage*

tions than the analyst is interested in. Alternatively, several calibration graphs covering smaller, overlapping concentration ranges can be constructed.

3.11.4 Bias, Leverage and Outliers

Points of influence, which may or may not be outliers, can have a significant effect on the regression model and therefore on its predictive ability. If a point is in the middle of the model (*i.e.* close to $\bar{x}$), but outlying on the y-axis, its effect will be to move the regression line up or down. The point is then said to have influence because it introduces an offset (or bias) in the predicted values (see Figure 3.13a). If the point is towards one of the extreme ends of the plot its effect will be to tilt the regression line. The point is then said to have high leverage because it acts like a lever and changes the slope of the regression model (see Figure 3.13b). Leverage can be a major problem if one or two data points are a long way from all the other points along the x-axis.

A leverage statistic (ranging between $1/n$ and 1) can be calculated for each value of x. There is no set value above which this leverage statistic indicates a point of influence. A value of 0.9 is, however, used by some statistical software packages.

$$\text{Leverage}_i = \frac{1}{n} + \frac{(x_i - \bar{x})^2}{\sum_{i=1}^{n}(x_i - \bar{x})^2} \qquad \text{(Eq. 3.30)}$$

where x_i is the x value for which the leverage statistic is to be calculated, n is the number of points in the calibration and $\bar{x}$ is the mean of all the x values in the calibration. The term $\sum_{i=1}^{n}(x_i - \bar{x})^2$ represents the sum of the squared differences between each x value and the mean of the x values.

To test if a data point (x_i, y_i) is an outlier (relative to the regression model) the following outlier test can be applied:

$$\text{Test value} = \frac{|\text{residual}_{\max}|}{\text{rsd}\sqrt{1 + \frac{1}{n} + \frac{(y_i - \bar{y})^2}{(n-1)s_y^2}}} \qquad \text{(Eq. 3.31)}$$

Table 3.6   *Outlier test for simple least squares linear regression*

Sample size (n)	95% confidence table-value	99% confidence table-value
5	1.74	1.75
6	1.93	1.98
7	2.08	2.17
8	2.20	2.23
9	2.29	2.44
10	2.37	2.55
12	2.49	2.70
14	2.58	2.82
16	2.66	2.92
18	2.72	3.00
20	2.77	3.06
25	2.88	3.25
30	2.96	3.36
35	3.02	3.40
40	3.08	3.43
45	3.12	3.47
50	3.16	3.51
60	3.23	3.57
70	3.29	3.62
80	3.33	3.68
90	3.37	3.73
100	3.41	3.78

where rsd is the residual standard deviation, s_y is the standard deviation of the y values, y_i is the y-value, n is the number of points, $\bar{y}$ is the mean of all the y values in the calibration and residual$_{max}$ is the largest residual value.

For example, the test value for a suspected outlier in Figure 3.11 is 1.78 and the critical value is 2.37 (Table 3.6 for 10 data points). Although the point appears extreme, it could reasonably be expected to arise by chance within the data set.

Further worked examples that describe the use of CRMs for some of the points raised in this chapter can be found in Chapters 4–6.

3.12   References

1. T. J. Farrant, *Practical Statistics for the Analytical Scientist, A Bench Guide*, Royal Society of Chemistry, Cambridge, 1997 (ISBN 0-85404-442-6).
2. *Statistics – Vocabulary and Symbols, Part 1: Probability and General Statistical Terms*, ISO 3534-1, 1993.
3. *Guide to the Expression of Uncertainty in Measurement*, ISO, 1993 (ISBN 92-67-10188-9).

The Use of CRMs for Instrument Calibration

4.1 Instrument Calibration

Modern instrumental methods of analysis offer a wide range of potential benefits, such as low detection limits, high specificity, good precision and automated sample throughput. However, in most cases the relationship between the output signal of an instrument (*e.g.* peak area, counts, mV, *etc.*) and the quantity of analyte being measured (*e.g.* g, mole, *etc.*) is empirical in nature. There is usually no well-understood physical or chemical theory that describes the magnitude of the signal in terms of the quantity of analyte present. Consequently, the amount of analyte present in a test sample cannot be determined from the measured instrument signal from first principles. The utility of most analytical instruments arises entirely from the experimental observation that the observed instrument signal is some arithmetical function of the quantity of analyte:[1]

$$\text{Signal} = K \times (\text{analyte quantity})^n \qquad \text{(Eq. 4.1)}$$

For the commonly encountered linear relationship between signal and analyte quantity, $n = 1$. The proportionality constant K is usually unknown, owing to the lack of an appropriate physical/chemical theory to underpin the fundamental operation of the instrument.

Under these circumstances, it is necessary to calibrate the output signal of the instrument by introducing accurately known quantities of the particular analyte(s) of interest. The signal thus obtained from the calibrant is then compared with the signal obtained from the test sample and the quantity of analyte in the sample is often determined by a calculation of the type:

$$\text{Analyte quantity in test sample} = \frac{\text{Sample signal}}{\text{Calibrant signal}} \times \text{Amount of analyte in calibrant}$$

$$\text{(Eq. 4.2)}$$

This calculation would apply where the instrument signal varied linearly with analyte quantity (*i.e.* $n = 1$ in Eq. 4.1). It will be readily apparent that the validity of the calculation represented by Eq. 4.2 depends on the values for n and K in Eq. 4.1 being the same for both the calibrant and the test sample. In other words, the instrument must respond quantitatively to the analyte in the same manner as to the analyte in the test sample. Only then will we be comparing like with like; in any other circumstance the comparison of the calibrant signal with the test sample signal will not be valid and an erroneous analytical result will be produced. Thus, we must establish that the experimental conditions used to calibrate an instrument are appropriate for the test samples that are to be analysed. The proper selection and use of appropriate reference materials as instrument calibration standards is therefore discussed in detail in this chapter.

4.2 Examples of CRMs Used for Instrument Calibration

4.2.1 A Note on Terminology

To clarify an issue of terminology, when the term *reference material* is used in the context of instrument calibration, the term is used to cover a number of different types of material, the main ones of which are:

- a pure substance, with a documented purity value
- a pure substance with a documented physical property, such as melting point
- a solution of a substance with a documented concentration value
- a solution of a substance with a documented property value, such as absorbance
- an homogeneous mixture of a substance in a solid matrix, with a documented concentration value
- an homogeneous mixture of a substance in a gaseous matrix, with a documented concentration value

Where the documented concentration value or property value has been established and certified by procedures for which the traceability and measurement uncertainty can be established (see Chapter 2), the material is described as a *certified* reference material. A further issue of terminology is that analysts frequently (and colloquially) describe CRMs that are used to calibrate instruments as *standards*, the term CRM being reserved in everyday speech for matrix substances used for checking the accuracy of *entire* analytical methods. For the purposes of this text, the terms *CRM* and *standard* are regarded as being synonymous.

The following provides some examples of the types of CRM used for instrument calibration.

4.2.2 Pure Substance with a Documented Purity Value

The CRM LGC1110 (*p,p'*-DDE) has a certified purity value of 99.6 mass/mass per cent. A typical use of this material would be for the calibration of GC–MS equipment used for the determination of trace amounts of organochlorine pesticide residues, such as *p,p'*-DDE, in foodstuffs and environmental samples. This application would normally require the analyst to prepare a standard solution by taking an accurately measured mass of the material and dissolving it in an accurately measured volume (or mass) of an appropriate organic solvent. The certificate for the material also documents the uncertainty of the purity value (±0.4%). With this information, and by establishing the uncertainties associated with the preparation of the solution, the analyst is able to calculate the uncertainty of the concentration of the standard solution prepared. This uncertainty may then be compared with the uncertainty expected in the result obtained for the trace analysis of *p,p'*-DDE in a food or environmental sample. The uncertainty in the standard solution should, of course, be small compared with the overall uncertainty, so that the calibration standard makes a negligible contribution to the overall uncertainty.

Similar CRMs would be samples of pure polyaromatic hydrocarbons (PAHs), polychlorinated biphenyls (PCBs), pharmaceuticals, metals, metal salts, *etc.*, all of which, provided they have a documented purity value, could be used to prepare calibration standards for a particular analysis.

4.2.3 Pure Substance with a Documented Physical Property

The CRM LGC2409 (carbazole) has a certified melting (liquefaction) point of 245.58 °C. An analyst could use this material to calibrate the temperature scale of apparatus used to determine the melting points of substances such as pharmaceuticals. Melting point is often included in the specifications of pharmaceuticals as one measure of the purity of the pharmaceutical in question. A comparison of the observed melting point with the specified value under the same conditions enables a decision to be made as to whether the substance under test is within specification. In making this decision, account has to be taken of the overall uncertainty in the calibration procedure. The latter can be estimated by combining the uncertainty quoted for the certified melting point (± 0.07 °C) with the analyst's estimate of the uncertainty of the measurement of the melting point of the certified material.

The CRM LGC2601 comprises a sample of high purity indium (99.99995%), with a certified enthalpy of fusion value (3.296 kJ mol) and a certified melting point (156.61 °C). This material is used for the calibration of the energy and temperature scales of differential scanning calorimeters. Such equipment finds application in the determination of the purity of organic compounds and in the study of the energy changes that accompany phase changes in a sample as the temperature of the sample is slowly increased.

The CRM SRM185 g, produced by the National Institute of Standards and Technology (NIST), USA, may also be considered as a material with a documented physical property. It consists of a sample of potassium hydrogen phthalate and is certified to produce a solution of pH 4.006 at 25 °C (using the NIST pH scale) when the solution is prepared according to the instructions given on the certificate. The material is used for calibrating the scales of commercial pH meters.

4.2.4 A Solution of a Substance with a Documented Concentration Value

The CRM LGC4105 consists of a solution of copper in approximately 1 mol L^{-1} nitric acid, with a certified copper concentration of 10000 mg kg^{-1}. The concentration is also certified on a volume basis and at 20 ± 2 °C, the certified concentration value is 10530 mg L^{-1}. One application of this solution, perhaps after appropriate dilution, is the calibration of instruments such as atomic absorption spectrophotometers and inductively coupled plasma optical emission spectrometers. They can then be used in the determination of the copper content of a variety of sample types, such as foodstuffs, drinking water, animal feeds, *etc.* Again, the certificate also documents the uncertainty of the certified concentration value ($\pm$ 60 mg kg^{-1}), enabling the analyst to estimate the contribution the calibration standard makes to the overall uncertainty of a copper determination when using this standard.

The CRM LGC5401 consists of a solution of ethanol in water, with a nominal ethanol concentration of 80 mg (100) mL^{-1}. The actual ethanol concentration is certified to two decimal places for each new batch of the solution produced and an uncertainty value is also quoted on the certificate. The solution is used for calibrating equipment used to determine ethanol in blood and breath samples that have been taken in connection with drink-driving legislation.

Similar CRMs, for example, are solutions of compounds such as pesticides, PCBs, PAHs, phenols, anions, *etc.*, with documented concentration values, which could be used to calibrate an instrument used in the determination of the particular species concerned.

4.2.5 A Solution of a Substance with a Documented Property Value

The CRM LGC2011 (Set No. S2 69) consists of a set of three cuvettes containing solutions of varying concentrations of mixtures of sodium nitrate, cobalt chloride and nickel chloride in 10% perchloric acid. The absorbance values of the cuvettes are certified at a number of wavelengths in the UV/visible range (Table 4.1).

These cuvettes are used to calibrate the absorbance scale of a UV/visible spectrophotometer in the range 0.1–1.5.

Similar CRMs are solutions with certified wavelengths of absorption in the UV/visible range (LGC2010) and solutions of polystyrene in hexane (LGC2012), with certified wavelength values for the IR region. Such certified

Table 4.1

Cuvette	Absorbance at 719.0 nm	Absorbance at 512.5 nm	Absorbance at 395.0 nm	Absorbance at 299.4 nm
A	0.618 ± 0.003	1.493 ± 0.004	1.502 ± 0.004	1.504 ± 0.004
B	0.372 ± 0.003	0.897 ± 0.003	0.914 ± 0.003	1.012 ± 0.003
C	0.125 ± 0.003	0.294 ± 0.003	0.300 ± 0.003	0.310 ± 0.003

solutions are used for calibrating the wavelength scales of UV/visible and IR spectrophotometers.

4.2.6 A Mixture of a Substance in a Solid Matrix with a Documented Concentration Value

The CRM SRM1216 (produced by NIST, USA) consists of a series of chemically modified micro-particulate silica, with certified levels of carbon in the range 0.7–17.04%. These materials are used for calibrating instruments that are used to determine the total elemental carbon content of samples. The uncertainties quoted for the certified values enable the uncertainty of the calibration to be calculated.

A further example is provided by SRM2579, which consists of a series of mylar sheets coated with a single uniform paint layer, the certified lead content of which is in the range 3.53 mg cm^{-3} to < 0.0001 mg cm^{-3}. This CRM is used for calibrating X-ray fluorescence (XRF) field units for the evaluation of leaded paint hazards in public housing. XRF is known to be particularly sensitive to matrix influences, so it is especially important to use CRMs as calibration standards whose matrix closely resembles that of the test samples.

4.2.7 A Mixture of a Substance in a Gaseous Matrix with a Documented Concentration Value

CRMs comprising mixtures of CO, CO_2 and propane in nitrogen are prepared and issued by the UK National Physical Laboratory (NPL). One application of these materials is the calibration of equipment used to monitor vehicle exhaust emissions for pollutants. The accuracy of all vehicle emission testing instruments is checked at three-monthly intervals, using gas mixtures traceable to the NPL standards.

Another CRM issued by NPL comprises a multi-component mixture of hydrocarbon standards, which is used to calibrate instruments for monitoring air for volatile organic compounds (VOC).

Calibration Standard	Sample
Obtain a suitable CRM for use as a calibration standard	Receive bulk sample
⇓	⇓
	Prepare bulk sample (*e.g.* dry, mix, grind, filter, *etc.*)
⇓	⇓
	Take sub-sample for analysis (by mass or volume)
⇓	⇓
Prepare calibration standard (*e.g.* dissolution, *etc.*)	Prepare sub-sample (*e.g.* dissolution, digestion, extraction, clean-up, dilution, make to volume, *etc.*)
⇓	⇓
Introduce calibration standard to instrument	Introduce sample solution to instrument
⇓	⇓
Make instrumental measurement of the calibration standard	Make instrumental measurement of the sample solution
⇓	⇓
Estimate the standard uncertainty of the calibration measurement, *u*(calibration)	Estimate the standard uncertainty of the sample measurement, *u*(sample)* solution
⇓	⇓

Compare instrument readings for the sample solution and the calibration standard

⇓

Calculate result for the concentration of the analyte in the sample and its uncertainty

*It should be appreciated that *u*(sample) is determined by the uncertainties arising from **all** steps of the analytical procedure that are applied to the sample and **not** just the instrumental measurement step.

Figure 4.1

4.3 Instrument Calibration as Part of the Analytical Process

To carry out an instrumental calibration procedure effectively, it should be appreciated that the calibration step is usually only one part of the entire analytical process that is applied to the test sample. This is illustrated by the flow chart shown in Figure 4.1 which lists, in sequence, the main operations that are likely to be involved in a typical analytical procedure, although not all analytical procedures will necessarily consist of all of these operations.

Some important consequences of the foregoing discussion are considered further below.

4.3.1 Sample Matrix Effects

Operations such as sampling, grinding, mixing, sub-sample weighing, addition of internal standard, sample dissolution, dilution to known volume, digestion, extraction, extract clean-up, *etc.*, often have to be carried out before the sample or its extract can be introduced to the instrument.

The effect of these operations on the physical and chemical form of the prepared sample that is ultimately subjected to instrumental measurement must be carefully considered. Appropriate steps must be taken to ensure that the physical and chemical form of the CRM used to calibrate the instrument is sufficiently similar to that of the sample. In this way, the instrument responds quantitatively to both sample and calibrant in the same manner, as discussed in Section 4.1 above. In the absence of this condition, 'matrix effects' will come into operation, leading to either a suppression or enhancement of the instrument signal for the analyte in the sample matrix compared with the signal observed for the same amount of analyte when present in the calibrant. Procedures to deal with matrix effects are discussed in Section 4.13.

4.3.2 The Accuracy Required in the Calibration Procedure

At the outset of any analysis, the analyst should have an awareness of the accuracy that is required in the result that is to be reported to the customer. The required accuracy will, of course, vary with the customer's intended application of the result. For example, the customer may need to enforce a contamination limit of 2 mg kg^{-1}. A typical measurement uncertainty of ±30% would not be acceptable where most samples are expected to fall only slightly below this limit, but could be accepted for a screening procedure where most of the samples will be well below the contamination limit. Also, the consequences of making a wrong decision due to the uncertainty in a measurement result must be considered. In some cases the consequences may be modest and the acceptable uncertainty in an analytical result can be correspondingly higher, compared with a situation where the consequences are more severe. For example, in a large geological survey, it may be important to gather many results quickly at low cost. The situation will be totally different for, say, clinical samples taken from individual patients, where the consequences of a wrong decision could be life threatening.

Once an estimate is made of the uncertainty that is acceptable in the analytical result, a decision can be made as to the uncertainty that is acceptable in the calibration of the analytical instrument being used. A key principle to remember is that the accuracy of an analytical result can never be better than the accuracy of the calibration on which that result is based. In practice, the uncertainty in a result will nearly always be greater than the calibration uncertainty. The following two examples illustrate an approach that may be taken to estimating the acceptable uncertainty in an instrumental calibration.

(a) The Determination of Trace Levels of Organochlorine Residues in Oil
If it is desired to determine the *p,p'*-DDE content of a sample of animal or vegetable oil at an expected level of 1 mg kg^{-1}, with a standard uncertainty of ±10%, the standard uncertainty associated with the calibration (*u*(calibration)) must be no greater than this figure. In practice, it must be considerably less,

Table 4.2

u(calibration) (%)	u(sample) (%)	u(overall) (%)
10	10	14.1
8	10	12.8
5	10	11.2
3	10	10.4
1	10	10.1

because the operations involved in the preparation and instrumental measurement of the sample will also contribute to the uncertainty of the analytical result. Indeed, experience shows that for this type of trace analysis, the standard uncertainty in a result due to the preparation and measurement of the sample (u(sample)) can typically be expected to approach ±10% at the 1 mg kg^{-1} level. Under these circumstances, we should endeavour to carry out the instrument calibration in a way that makes a negligible contribution to the overall measurement uncertainty in the final result.

We can consider the overall standard uncertainty of an analytical result to be composed as follows:

$$u(\text{overall}) = \sqrt{\{u(\text{calibration})^2 + u(\text{sample})^2\}} \qquad \text{(Eq. 4.3)}$$

From this equation, Table 4.2 may be constructed.

It is readily seen that if the standard uncertainty due to instrument calibration is less than 0.3 of the standard uncertainty due to the sample preparation and measurement steps, then the calibration will make only a minor contribution to the overall measurement uncertainty and can, for all practical purposes, be ignored.

In the example of p,p'-DDE in oil, we can conclude that the standard uncertainty due to calibration should not exceed 3%; it will then make no significant contribution to the target standard uncertainty of 10%. Under these circumstances, we can tolerate a sample preparation procedure with a standard uncertainty of 10% or less.

In contrast, if we were confident that the sample preparation steps would contribute a standard uncertainty of not more than 5% to the final result, we could tolerate an instrument calibration step with a standard uncertainty of 8.5% and still meet the requirement of producing a result with an overall standard uncertainty of 10%.

(b) The Determination of Ethanol Levels in Blood

In the UK, the legal limit for a driver's blood ethanol level is 80 mg (100 mL)$^{-1}$, although prosecutions are only initiated when the measured level exceeds 87 mg (100 mL)$^{-1}$. The tolerance of 7 mg (100 mL)$^{-1}$ is based on the expected measurement uncertainty associated with this determination. The standard uncertainty for the measurement is estimated to be 2.3 mg (100 mL)$^{-1}$, thus a measured value of 87 mg $(100 \text{ mL})^{-1}$ can be held, with 99% confidence, to exceed the permitted maximum of 80 mg $(100 \text{ mL})^{-1}$.

If a standard uncertainty of 2.3 mg $(100 \text{ mL})^{-1}$ for a measurement of alcohol in blood is to be attained, it is clear that the standard uncertainty due to the calibration of the equipment used must be less than this. Following similar reasoning to that above, if the standard uncertainty in the calibration is less than $0.3 \times 2.3 = 0.7$ mg $(100 \text{ mL})^{-1}$, then the calibration will make no significant contribution to the overall standard uncertainty of the ethanol in blood result. In this case, therefore, the standard uncertainty of the calibration must be less than 0.8% of the final reported result, a considerably more stringent requirement than in the example of trace analysis given in (a) above.

The practical factors affecting the uncertainties that are associated with instrument calibration operations are discussed further in Section 4.4.2.

4.4 Choosing a CRM for Use as a Calibration Standard

When selecting a CRM for use as an instrument calibration standard a number of factors will need to be considered.

4.4.1 Documentation for the CRM

As all analytical data are ultimately traceable to the calibration standard on which the data are based, it is important that the CRM used as the calibration standard is as fully documented as possible. The producer of the CRM will issue a certificate to accompany it which should be retained as an important part of the analytical records of the laboratory using the CRM. A certificate for a CRM will usually include the information listed below. In the absence of a certificate, the user should endeavour to establish as much information as possible about the material, using the list as a guide.

1. Name and address of the producer of the material
2. Description/name of the material
3. Physical/chemical form of the material
4. Sample number/batch number/certificate number
5. Date of production/issue
6. Shelf life/expiry date
7. Property values
8. Uncertainty of property values
9. Procedures used to characterise the material (determine the property value)
10. Information on the traceability of property values to the SI (kg, metre, mole *etc.*)

Further aspects of some of these features of are discussed in Sections 4.4.2 to 4.4.5 below.

4.4.2   The Magnitude and Uncertainty of the Property Values

The magnitude of the property value (*e.g.* concentration) of the CRM used as the calibration standard must be appropriate to the sample being analysed. For example, if we are determining the nitrate level in a sample of drinking water, where the expected level is 50 mg L^{-1}, by ion chromatography, then the concentration of the calibration standard needs to be a close approximation to this nominal value. It would not be appropriate to calibrate the instrument using a calibration standard with a concentration of, say, 5 or 500 mg L^{-1}.

As discussed in Section 4.3.2, the accuracy of an analytical result on a sample can never be any better than the accuracy of the calibration procedure on which the result is based. It was noted in Section 4.3.2 that where the standard uncertainty of the calibration procedure is less than 0.3 of the target standard uncertainty for the result (or less than 0.3 of the standard uncertainty of the result due to the sample preparation and measurement steps of the analytical procedure), then the contribution of the calibration standard uncertainty to the overall standard uncertainty of the result may be regarded as being insignificant.

It is important to note, however, that the calibration uncertainty arises from three main sources:

- the uncertainty of the documented property value of the CRM used as the calibration standard [$u(\text{CRM})$]
- the uncertainty associated with any preparation of the CRM prior to its introduction to the instrument [$u(\text{CRM}_{\text{preparation}})$]
- the uncertainty associated with the measurement of the calibration standard by the instrument [$u(\text{CRM}_{\text{measurement}})$]

For example, the CRM LGC1110 (*p,p'*-DDE) has a certified purity value of 99.6 mass/mass percent, which has an (expanded) uncertainty ($k = 2$) of 0.4% (see Section 3.10). As discussed in Section 4.2.2, a typical use of this material would be to calibrate GC–MS equipment, which would require the analyst to prepare a standard solution of the *p,p'*-DDE in a suitable solvent. The fact that the documented standard uncertainty of the purity value is 0.2% indicates that the standard uncertainty associated with the instrument calibration procedure will be at least this amount. In practice it will, of course, be greater owing to the other two contributing factors listed above. The analyst will need, therefore, to estimate the standard uncertainty due to preparation and measurement of the calibration standard and then combine this with the standard uncertainty of the purity of the CRM in order to estimate the total standard uncertainty of the calibration process. This can be described by the equation:

$$u(\text{calibration}) = \sqrt{\{u(\text{CRM})^2 + u(\text{CRM}_{\text{preparation}})^2 + u(\text{CRM}_{\text{measurement}})^2\}} \quad \text{(Eq. 4.4)}$$

If the value of $u(\text{calibration})$ is 0.3 (or less) of the target standard uncertainty for the final result, its contribution to the target standard uncertainty can be ignored.

It is seen from Eq. 4.4 that the standard uncertainty of a CRM used as a calibration standard will, in its turn, make a negligible contribution to the *total* standard uncertainty of the instrument calibration step, provided it is less than 0.3 of the standard uncertainty due to the preparation and measurement of the calibration standard. As a 'rule of thumb', therefore, it follows from this that if the uncertainty in the documented value of a CRM used as a calibration standard is less than 0.1 of the target uncertainty for the final result, its contribution to the overall standard uncertainty may be regarded as being negligible.

By way of example, the p,p'-DDE CRM will make a negligible contribution to the uncertainty of a measurement where the overall target standard uncertainty is 2% or more. If the required target standard uncertainty is, say, 1%, two options are available to the analyst:

- use a different p,p'-DDE CRM with a lower standard uncertainty (0.1% or less)
- try to reduce the standard uncertainty due to the preparation and measurement of the calibration standard

However, when considering these options, it is important to remember that the contribution to the overall target standard uncertainty from the preparation and measurement of the sample is often likely to be higher than the contribution from the instrument calibration step. Under these circumstances, there is no benefit to be gained by reducing the standard uncertainty of the calibration procedure beyond the point where it is less than 0.3 of the standard uncertainty due to analytical operations carried out on the sample.

4.4.3 How the Property Value Is Characterised

The property value of a CRM used as a calibration standard may have been characterised by the producer of the material in one of a number of possible ways (see Chapter 2). These include:

1. the use of a primary (definitive) measurement method
2. on the basis of the accurate gravimetric or volumetric preparation of the material
3. the use of two or more independent measurement methods
4. from data obtained in an inter-laboratory study of the material
5. the use of a single measurement method

The user should ensure that the property value has been established by a means that is sufficiently rigorous for the user's intended application of the CRM. In general terms, CRMs whose property values have been characterised by procedures 1–4 comply, in principle, with the requirements for the characterisation of certified reference materials described in ISO Guide 35.[2] As such, the documented property value may be regarded, for all practical purposes, to be an unbiased estimated of the true value. Such procedures also allow an estimate to be made of the uncertainty of the property value and analysts should ensure, wherever possible, that the materials they are using have documented uncertainty values. Reference materials of this type are often described as *certified* reference materials, as opposed to reference materials (*i.e.* non-certified materials), which have been characterised in a less rigorous manner. As a consequence, the uncertainty associated with the property value of a reference material is likely to be higher than that for a certified reference material. Indeed, for reference materials an uncertainty estimate for the property value may not always be available. The use of CRMs is therefore preferred for instrument calibration operations.

However, due to lack of availability of suitable CRMs, the use of a reference material for calibration purposes may be necessary. Under these circumstances, the analyst should endeavour to ensure that the chosen reference material has some documentation describing its property value and how the property value was determined (see Section 4.4.1). Examples of reference materials would be those characterised by Method 5 above.

Both certified reference materials and reference materials may be characterised by Method 5, in which the property values are established by means of a single measurement technique. An example would be the characterisation of the fat content of a foodstuff by a precisely specified analytical method that, by definition, determines fat. Another example would be the determination of leachable metals (as opposed to total metals) in a soil material, where the leaching conditions are precisely defined. In both of these cases, the documented property values of the matrix reference materials only apply to the specific measurement procedure concerned and users should ensure that they are following the defined method exactly. Such materials are, in fact, encountered quite frequently in analytical chemistry.

A different type of Method 5 material would be, for example, a sample of an organic compound such as the antioxidant butylated hydroxyanisole (BHA). This might well be used as a calibration standard in an HPLC method to determine BHA in frying oil. The compound BHA may not be available as a certified reference material, with a *certified* purity value, but commercially produced BHA, with a manufacturer's reported purity value (but not an uncertainty value) will be available. Depending on the quality of the information available, the analyst should consider using some procedure to verify the manufacturer's reported purity value. Perhaps a purity estimate by differential scanning calorimetry, gas–liquid chromatography or high performance liquid chromatography could be obtained. Alternatively, a second sample of the material could be obtained *from an independent source* and then compared with the original sample of the reference material.

4.4.4 The Traceability of the Property Value

Where a CRM is available, the certificate may contain a statement on the traceability of the certified property value. If the property value has been established by a definitive measurement procedure or by an accurate gravimetric/volumetric formulation procedure, the property value may be said to be traceable to the base units of the SI system such as the kilogram, the metre or the mole. Often, however, traceability of this fundamental type will not be realisable. Under these circumstances, the calibration of the instrument is regarded as being traceable to the CRM, rather than to the SI. However, as the CRM will have a certificate, issued by a recognised producer, the calibration will nevertheless be traceable to a measurement standard (in this case, the CRM itself) that is recognised by (and available to) the analytical community at large.

It will be appreciated, of course, that the use of a reference material (*i.e.* a non-certified material) only allows traceability of the calibration to be established to the reference material. However, the value of such traceability is questionable, especially as different laboratories may well be using different sources of a particular reference material, so that there is no common, recognised measurement standard in use. In these circumstances, an observed variation in results when different laboratories analyse the same sample is often due to the fact that different calibration standards, of variable quality, are being used.

4.4.5 Physical/Chemical Form – Pure Substance or Matrix Material?

CRMs will usually be available as pure substances (solids, liquids and gases), solutions of pure substances (either single component or multi-component) in an aqueous or organic solvent or mixtures in either a solid or gas matrix.

When the calibration standard is introduced to the instrument, it must ideally be in a physical and chemical form that is sufficiently similar to the final form of the prepared test sample so that any errors in the analytical result due to matrix effects (see Section 4.3.1) are negligible. If this is not possible, then a calibration procedure to compensate for matrix effects must be applied (see Section 4.13).

For a determination of, for example, trace metals in foodstuffs by ICP–OES the most likely choice for an instrument calibration standard will be between a sample of the pure metal with a documented purity value and a commercially available solution of the metal, with a documented concentration value. In the former instance, the analyst would have to prepare a calibration standard solution by dissolving an accurately measured mass of the metal in acid solution and making the solution to an accurately known mass or volume. From a metrological point of view, there is little to choose between these two options, although a commercially available solution may be more convenient for most laboratories, especially where the uncertainty of the concentration value is provided with the solution.

Although CRMs comprising foodstuffs with documented levels of trace metals are available, these materials are **not** recommended for instrument calibration purposes, as the uncertainty of the documented values is significantly greater than the uncertainties associated with pure metals and their solutions. For example the CRM LGC7000 is a beef/pork matrix with a certified zinc concentration of 14.2 mg kg^{-1}, with a standard uncertainty of 0.2 mg kg^{-1}, which is 1.4% of the certified value. In contrast, the CRM LGC4120 is a solution of zinc in 1 mol L^{-1} nitric acid, with a certified concentration of 1000 mg kg^{-1}, with a standard uncertainty of 3 mg kg^{-1}, which is 0.3% of the certified value. Thus, an instrument calibration with this material would have a significantly lower uncertainty by at least a factor of five. Furthermore, the beef/pork matrix may well affect the quantitative response of the ICP–OES instrument, so that the calibration obtained is only strictly applicable to a sample with a matrix of the same type. Yet a further consideration is that the preparation of the matrix sample by the user laboratory involves such steps as sample mixing (to homogenise the matrix) and sample digestion. These operations are inherently subject to greater experimental errors than the more straightforward preparation of a standard metal solution.

For reasons of this type, instrument calibration using a CRM comprising a pure substance (or its solution in a suitable solvent) is generally the procedure of choice for most analytical measurements. If the matrix of the sample being analysed is thought to modify the instrument response compared to that of the calibration standard, then the procedures used to deal with matrix effects (see Section 4.13) are used in conjunction with the pure substance calibration standard.

However, all of the above considerations notwithstanding, there may be situations where an instrument must be calibrated with a matrix CRM. Such an instance would arise where the procedures normally used to deal with matrix effects are insufficient to provide appropriate allowance for the effect of the sample matrix on the instrument response. For example, it is widely recognised that the technique of X-ray fluorescence (XRF) for the quantitative determination of major and trace metals in metallurgical, geological and other inorganic samples is acutely dependent on their matrices. Elements in the matrix markedly influence both the absorption of the incident X-rays and the intensity of the fluorescent X-rays emitted by the particular target element being measured. One approach widely used is to calibrate the XRF instrument with a range of CRMs similar in composition to the sample being tested and with documented levels of the target analyte. As an example, NIST produces a range of CRMs (SRM1762 to 1768) that comprise a low alloy steel matrix, with certified levels of elements such as C, Mn, P, S, Si, Cu, Ni, Cr, *etc.* This approach requires, of course, that the approximate composition of the test sample is known, a situation that is not uncommon in industrial control analyses. Where the composition of the sample and standards is very similar, the intensity of fluorescent X-rays from the sample can be compared directly (linearly) with the observed intensity from the calibration standards.

Another example of an instrumental analytical technique which utilises calibration with matrix CRMs is atomic emission spectroscopy, where the target analyte is determined by measurement of the intensity of its atomic emission, after excitation by such means as an AC or DC arc, or a spark discharge.

4.4.6 Storage Requirements and Shelf Life/Expiry Date

When selecting and using a CRM for use as a calibration standard, attention needs to be paid to the physical and chemical stability of the material. For a CRM, information should be provided on the certificate and it is important that any special storage requirements are observed. Thus, it may be necessary to store the material in the dark, in a desiccator, in the absence of air, at sub-ambient temperatures (*e.g.* 4 °C, -20 °C, -70 °C), *etc*. Users should ensure that they have the necessary facilities for the proper storage of the CRM, otherwise the documented property values may become invalid.

Another important consideration during storage (and use) of a CRM is the need to scrupulously avoid contamination of the material. If this occurs, the documented property values will again be rendered invalid. Such contamination can take the form of moisture pick-up and surface oxidation, as well as inadvertent chemical contamination by storage adjacent to incompatible chemicals or careless dispensing of the material using contaminated glassware and spatulas. Where a solution is prepared from the CRM, care should be taken to store the solution properly, in order that factors such as solvent evaporation do not alter the prepared concentration value.

It is also important to observe any expiry date quoted by the producer of the CRM, as the documented property values cannot be guaranteed to be valid after this date.

4.5 The Preparation of a Calibration Solution Using a CRM

Some preparation of the CRM is often required before it can be introduced to the instrument and frequently the preparation of a solution of the calibration standard is necessary. Such operations will, in principle, add to the uncertainty of the documented property value, as described in Section 4.4.2 above. The following example shows how the preparation of a solution can affect the uncertainty of the property value.

The CRM LGC 1110 (*p,p'*-DDE) has a certified purity value of 99.6% and an expanded uncertainty (coverage factor, $k = 2$) of 0.4%. A solution of the material, in hexane, with a nominal concentration of 0.1 mg mL^{-1} was prepared. Table 4.3 shows the values for the purity of the CRM, the mass of the CRM taken and the volume of hexane solvent used and their estimated standard uncertainties.

The concentration of the prepared solution is calculated to be 0.1135 mg mL^{-1}. The standard uncertainty of this value, calculated by combining the individual uncertainties in the usual manner[3] (*i.e.* as the sum of the squares of the individual relative uncertainties, ru^2 – see also Chapter 3, Section 10), is seen to be 0.0020 mg mL^{-1}. This is equivalent to 1.76% of the concentration value. Thus, the use of this solution for an instrument calibration would result in a calibration with a standard uncertainty of at least this value. It is of interest to note from the column '*ru*' in Table 4.3 that the mass value makes

Table 4.3 *Preparation of a calibration solution using a CRM (1)*

Source of uncertainty	Units	Value	u	ru	ru^2
Certified purity		0.996	0.002	0.002008	4.03E−06
Mass of CRM taken	g	0.0114	0.0002	0.017544	0.000308
Volume of hexane	mL	100	0.06	0.0006	3.6E−07
Concentration	mg mL^{-1}	0.1135	0.0020	0.017669	0.000312

u = standard uncertainty; ru = relative standard uncertainty ($= u$/value); ru^2 = relative standard uncertainty squared

by far the largest contribution to the standard uncertainty of the concentration value. The original CRM has a standard uncertainty that is only 0.20% of the certified purity value. Thus, the preparation of the solution has introduced significant additional uncertainty over and above that due to the CRM as issued by the producer. The analyst would need to decide whether the standard uncertainty of the prepared solution of 1.76% was acceptable for a particular analytical application (see also Sections 4.3.2 and 4.4.2).

If the standard uncertainty of 1.76% were not acceptable, the solution would need to be prepared in a different manner. One simple approach would be to carry out the weighing of the CRM using a five-figure, rather than a four-figure, balance. The uncertainty associated with the mass value of 0.01143 g is then estimated to be 0.00004 g, which leads to a standard uncertainty of 0.00046 mg mL^{-1} for the solution concentration, which is 0.40% of the concentration value, as shown in Table 4.4.

Although inspection of Tables 4.3 and 4.4 suggests that the measured volume of the hexane solvent is a negligible contributor to the standard uncertainty of the solution concentration, it will be realised, on further consideration, that the ambient temperature may influence the solution volume and therefore its concentration. For example, if the solution was prepared at 20 °C, but subsequently used in a laboratory whose temperature is 25 °C, the actual concentration of the calibrant introduced to the instrument will be less than the prepared value (0.1135 mg mL^{-1} in this example). It can be calculated that 100.0 mL of hexane at 20 °C, will occupy 99.3 mL at 15 °C and 100.7 mL at 25 °C. Thus, if

Table 4.4 *Preparation of a calibration solution using a CRM (2)*

Source of uncertainty	Units	Value	u	ru	ru^2
Certified purity		0.996	0.002	0.002008	4.03E−06
Mass of CRM taken	g	0.01143	0.00004	0.0035	1.22E−05
Volume of hexane	mL	100	0.06	0.0006	3.6E−07
Concentration	mg mL^{-1}	0.11384	0.00046	0.004079	1.66E−05

u = standard uncertainty; ru = relative standard uncertainty ($= u$/value); ru^2 = relative standard uncertainty squared

Table 4.5 *Preparation of a calibration solution using a CRM (3)*

Source of uncertainty	Units	Value	u	ru	ru^2
Certified purity		0.996	0.002	0.002008	4.03E−06
Mass of CRM taken	g	0.01143	0.00004	0.0035	1.22E−05
Volume of hexane	mL	100	0.4	0.004	0.000016
Concentration	mg mL^{-1}	0.11384	0.00065	0.005681	3.23E−05

u = standard uncertainty; ru = relative standard uncertainty ($= u$/value); ru^2 = relative standard uncertainty squared

we are confident that we can control the temperature of the calibration solution during both preparation and use to 20 ± 5 °C, then the standard uncertainty in the hexane volume may be estimated (assuming a rectangular distribution[3]) as $0.7/\sqrt{3} = 0.40$ mL. Comparison of this estimate for the standard uncertainty of the hexane volume with that given in Table 4.3 of 0.06 mL shows the influence of temperature effects on standard solutions in organic solvents. Table 4.5 shows the effect on the uncertainty of the concentration value. Thus, if there are variations in the ambient temperature of ± 5 °C during the preparation and use of the calibration solution, the standard uncertainty of the concentration value increases from 0.00046 mg mL^{-1} to 0.00065 mg mL^{-1}.

The general conclusion from the above discussion is that the analyst needs to be aware and take account of any aspects of the procedure used to prepare the calibration solution that may adversely affect the documented properties of the original CRM.

It should be appreciated that the above considerations are entirely different from the situation where the solvent evaporates from the solution, causing a gradual increase in the concentration value. The introduction of uncontrolled systematic errors of this type must be avoided by proper storage and possibly by appropriate monitoring (*e.g.* total mass measurements of the solution plus container) of the calibration solution.

Quite often a 'stock' calibration solution will be prepared and from this a number of more dilute solutions will be prepared, in order to cover the range of analyte concentrations of interest. Such dilutions will contribute to the uncertainty of the (diluted) concentration value. Table 4.6 illustrates the nature of these effects for the situation where an aliquot volume of 1 mL of the stock solution, measured by pipette, is diluted to 10 mL in a volumetric flask.

Inspection of the column '*ru*' in Table 4.6 shows that the standard uncertainty of the diluted solution is 0.7% of the concentration value, compared with 0.4% for the undiluted stock solution. Analysts need to consider the significance of such effects for the particular analysis they intend to carry out, since (as discussed in Section 4.3.2) the analytical result on a sample cannot be any better (and will be probably be worse) than the accuracy of the calibration standard. Furthermore, the use of the statistical technique of

Table 4.6 *Preparation of a calibration solution using a CRM (4)*

Source of uncertainty	Units	Value	u	ru	ru^2
Stock solution	mg mL^{-1}	0.11384	0.00046	0.004041	1.63E$-$05
Aliquot volume (pipette)	mL	1	0.0056	0.0056	3.14E$-$05
Final volume (flask)	mL	10	0.0144	0.00144	2.07E$-$06
Diluted concentration	mg mL^{-1}	0.011384	8.03E$-$05	0.007054	4.98E$-$05

u = standard uncertainty; ru = relative standard uncertainty ($= u$/value); ru^2 = relative standard uncertainty squared

least squares linear regression to construct the best straight line through a set of calibration points (see Section 4.8) depends on the assumption that the uncertainty in the concentration values of the calibration standards is insignificant compared to the uncertainty in the measured instrument signal for those standards. Knowledge of the expected uncertainties in the calibration solution concentrations allows the analyst to assess the validity of this assumption for a particular set of calibration data.

If the uncertainties in the prepared calibration solutions are considered to be too high, one option would be to prepare the solutions entirely by mass. Table 4.7 shows the calculated uncertainties when the stock solution was prepared by taking a nominal 0.1 g of the CRM and dissolving it in a nominal 66 g of hexane ($\approx$100 mL). A nominal 0.66 g aliquot of the stock solution was then diluted to a nominal 66 g with hexane. It is seen from the column 'ru' in Table 4.7 that the standard uncertainty of the diluted calibration solution is 0.20% of the concentration value. This compares with a standard uncertainty that was 0.7% of the concentration value when the dilution was carried out by volume. It may also be noted that for the dilution by mass, the standard uncertainty of 0.20% is due entirely to the standard uncertainty of the certified purity value of the CRM itself. Thus, by preparing the calibration solution in the manner shown in Table

Table 4.7 *Preparation of a calibration solution using a CRM (5)*

Source of uncertainty	Units	Value	u	ru	ru^2
Certified purity		0.996	0.002	0.002008	4.03E$-$06
Mass of CRM taken	g	0.11431	0.00004	0.00035	1.22E$-$07
Mass of hexane	g	66.1054	0.0005	7.56E$-$06	5.72E$-$11
Stock concentration	mg g^{-1}	1.7223	0.00351	0.002038	4.15E$-$06
Aliquot mass	g	0.65892	0.00004	6.07E$-$05	3.69E$-$09
Final mass	g	65.9046	0.0005	7.59E$-$06	5.76E$-$11
Diluted concentration	mg g^{-1}	0.01722	3.51E$-$05	0.002039	4.16E$-$06

u = standard uncertainty; ru = relative standard uncertainty ($= u$/value); ru^2 = relative standard uncertainty squared

4.7, no significant addition to the uncertainty of the original CRM has been incurred.

4.6 Calibration Models

By calibration model[1] is meant a mathematical description of the way the instrument signal varies with changing quantities (*e.g.* mass, concentration) of the particular analyte being measured. A number of models are observed in practice, the most important ones of which are described below.

4.6.1 Calibration Model 1

A commonly encountered model is where the instrument signal is a linear function of the quantity of analyte, which may be described by the equation:

$$\text{Signal} = K \times (\text{analyte quantity}) + B \qquad \text{(Eq. 4.5)}$$

K is a proportionality constant, the value of which does not need to be formally known or calculated, although it enters implicitly into the calculation of the analyte concentration in a sample when the instrument signal for the sample is compared with the instrument signal for the calibration standards. The value B represents the instrument signal obtained when a 'blank' calibration standard is measured (*e.g.* a solution identical to that used to prepare the calibration standards, but not containing the specific reference analyte of interest). Figure 4.2 shows a typical graph of a linear calibration model according to Eq. 4.5, for a value of $B = 0$ (*i.e.* where the instrument response to the calibration blank [analyte quantity = 0] is zero).

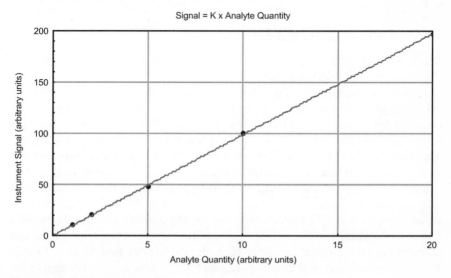

Figure 4.2 *Calibration model 1*

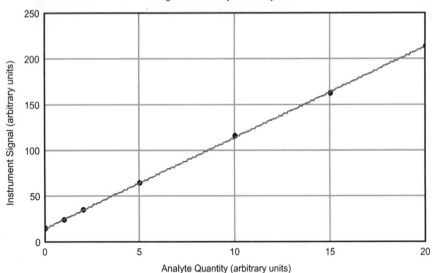

Figure 4.3 *Calibration model 2*

4.6.2 Calibration Model 2

A model that is sometimes encountered is where the instrument gives a signal even when a calibration blank is measured. Figure 4.3 shows the graph for calibration model 2, where B (the intercept of the graph) has a finite value that represents the instrument signal obtained when a calibration standard with a nominal reference analyte concentration of zero is introduced to the instrument.

4.6.3 Calibration Model 3

Figure 4.4 shows the graph for an important variation of calibration model 1 that is frequently observed in practice.

There is no simple equation for this model as it represents the situation where a linear relation between instrument signal and analyte quantity is observed for analyte quantities at the lower end of the range, but where the signal starts to plateau at higher analyte quantities. Such situations are not uncommon in analytical chemistry. For example, an electron capture detector which is used to quantify organochlorine compounds eluting from a GC column will enter the plateau region at analyte quantities that are sufficiently high to capture so many thermal electrons that additional organochlorine molecules have a reduced probability of effecting electron capture themselves. Ultimately, when the probability of additional organochlorine molecules effecting electron capture becomes zero, the detector is said to be saturated and the response curve becomes parallel to the x-axis and is not usable for quantitative work.

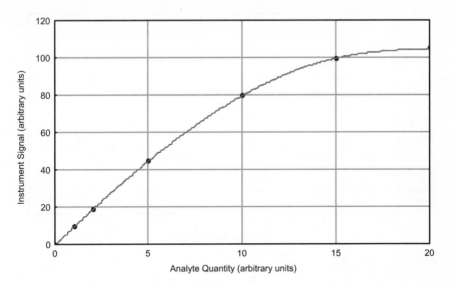

Figure 4.4 *Calibration model 3*

In the field of UV/visible spectrophotometry, a graph of absorbance *versus* concentration often exhibits a plateau at higher absorbance values. This may be due to a variety of reasons, either instrumental or originating in the solutions being measured. The technique of atomic absorption spectrophotometry may also exhibit a similar phenomenon.

4.6.4 Calibration Model 4

Some calibration graphs conform to a curvilinear model. One such example is the signal response of sulfur to a flame photometric detector, which follows an equation of the type:

$$\text{Signal} = K \times (\text{analyte quantity})^2 \qquad \text{(Eq. 4.6)}$$

Figure 4.5 shows a typical plot for this equation.

When confronted with this type of calibration model it is often more convenient to transform it to a linear model, by taking logarithms, viz.:

$$\text{Log (Signal)} = 2K \times \text{Log (analyte quantity)} \qquad \text{(Eq. 4.7)}$$

Figure 4.6 shows a typical plot for this equation.

A more general equation for the curvilinear model is given by a polynomial equation of the type:

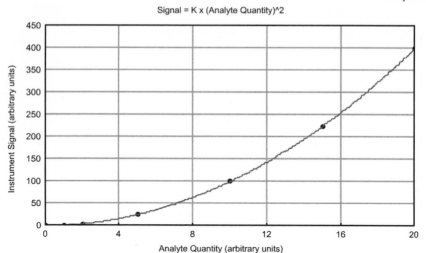

Figure 4.5   *Calibration model 4*

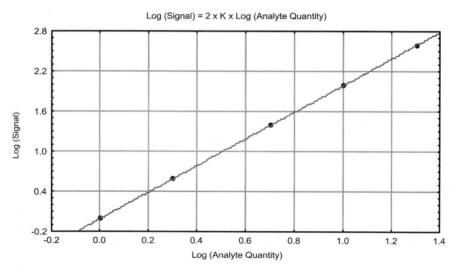

Figure 4.6   *Calibration model 4a*

$$\text{Signal} = K1 \times (\text{AQ}) + K2 \times (\text{AQ})^2 + K3 \times (\text{AQ})^3 + K4 \times (\text{AQ})^4 \ldots + B \quad (\text{Eq. 4.8})$$

where AQ is the quantity of analyte.

As a word of caution, it is unlikely in chemistry that a function higher than a quadratic (order 2) would be fitted to instrument calibration data. Additionally, there should be a scientific reason for expecting non-linearity before fitting of the data to a non-linear calibration model is considered.

In the following sections, a detailed procedure for investigating the linear calibration model is discussed.

4.7 Experimental Investigation of the Linear Calibration Model

The analyst should carry out a full investigation of the calibration procedure in circumstances such as setting up a new instrument, using an existing instrument for a new application, or where new types of sample are to be analysed using an existing instrumental procedure. Additionally, an established calibration model for an instrument routinely used on a particular application should be checked at regular intervals, perhaps weekly, monthly, three-monthly, *etc.*, depending on prior experience of the instrument.

The full calibration procedure will establish instrument parameters such as the following:

- linear range
- dynamic range
- sensitivity
- linear correlation coefficient
- calibration uncertainty

Whereas the linear range is that where the signal is a linear function of the analyte quantity, the dynamic range is that range over which there is at least a measurable change in signal with a change in analyte quantity. Use of the dynamic range effectively implies working in that region of the calibration curve where the graph starts to plateau (see Figure 4.3). The sensitivity of the instrument at a particular analyte concentration is represented by the gradient of the slope of the calibration graph at that concentration. Thus, sensitivity is constant within the linear portion of the calibration graph, but progressively decreases as the calibration line progressively approaches the horizontal. When the calibration line becomes horizontal the sensitivity is, of course, zero and the instrument is no longer usable for quantitative work. The instrument factor that influences the calibration uncertainty is the repeatability of the measurements made on the calibration standards. This source of uncertainty, combined with those due to the reference material itself and any preparation carried out on the reference material before it is introduced to the instrument, constitutes the total uncertainty of the calibration procedure (see Section 4.4.2).

Some important practical aspects[4,5] of the calibration procedure are discussed below.

4.7.1 Experimental Conditions

The operational parameters for the instrument should be those that will apply during the use of the instrument to analyse routine samples. Also, if several operators use the instrument routinely, they should all be represented in the calibration experiment.

4.7.2 Instrument Stability

It should be established that the instrument signal is not subject to drift with time, at least for the period of time required to carry out the measurements on the calibration standards and the measurements on the samples. If the drift is such that this cannot be guaranteed and no remedy can be found for the problem, then, rather than constructing a full calibration graph, the technique of bracketing could be used (see Section 4.12). The drift of the instrument may be assessed by replicate measurements of one particular calibration standard over a period of time. The mean of four replicate measurements of the calibration standard, carried out in succession over the shortest practical time scale, could be compared with that for similar measurements carried out at known, longer, time intervals. A *t*-test (see Chapter 3, Section 3.8.2) may be used to compare the statistical significance of any observed difference between the two mean results.

4.7.3 Choice of CRMs

General principles for the selection of a suitable CRM(s) for use as a calibration standard are discussed in Section 4.4.

4.7.4 Range of Property Values to be Covered

The property values (*e.g.* the concentrations of the reference analyte) of the calibration standards prepared from the selected CRM(s) must cover the range of values encountered in the analysis of routine samples. Where the calibration is to be carried out using a solution of a pure substance, a series of solutions is likely to be prepared by dilutions of a stock solution (see Section 4.5). However, it should be realised that in these circumstances the diluted solutions are not truly independent of the stock solution. *Ideally*, the individual calibration solutions should be prepared from independent weighings of the original pure substance CRM. Where matrix CRMs are used (see Section 4.4.5), the documented property values should, ideally, have been established independently of one another.

It is recognised, however, that because of considerations such as time, cost and lack of availability of suitable CRMs, the ideal situation will not always be attainable. The analyst should then endeavour to carry out an instrument calibration in a manner that approximates to the ideal as far as is considered to be practicable and justifiable, given the analysis concerned and the intended ultimate application of the analytical result (see also Section 4.3.2).

4.7.5 Number of Calibration Points to be Obtained

In order to establish the calibration line with an acceptable uncertainty and in a manner that is cost-effective, it is recommended that seven calibration standards covering the range of interest are obtained and/or prepared.

4.7.6 Spacing of the Calibration Points

One of the calibration standards should have a property value that is close to the centre of the values expected from the samples that are to be analysed. Two calibration standards should be prepared with property values that are similar to the highest and lowest values expected from the test samples. It might also be a useful precaution to prepare a high calibration standard that is somewhat larger than the expected sample value and a low calibration standard that is somewhat smaller than the expected sample value. The remaining two calibration standards should have property values that are spaced approximately equidistantly between the centre points and the extreme points.

For example, if samples of drinking water were being analysed for iron, where the expected concentrations lie within the range 10–50 μg L^{-1} (*i.e.* a central value of 30, the following calibration standards could be prepared: 5, 10, 20, 30, 40, 50, 60 μg L^{-1}.

A calibration line prepared in this way will have a mid-point that is close to the analyte level expected in the samples being analysed. This is the ideal situation, as the uncertainty of the calibration line is at a minimum at the mid-point of the line. The calibration uncertainty progressively increases as the upper and lower points of the calibration line are approached (see Section 4.8).

4.7.7 Number of Replicate Measurements

It is recommended that each calibration standard be measured in four-fold replicate. *Ideally*, where calibration solutions have been prepared, the four measurements should be obtained on four *different* solutions, *independently* prepared to have the same concentration of the reference analyte. In this way a true and full estimate of the random variation of the calibration measurement is obtained. The alternative approach of carrying out four replicate measurements on a *single* calibration standard will only give a partial assessment of the calibration variation, namely that due to the repeatability of the instrument readings only.

4.7.8 Treatment of Results

Table 4.8 shows a typical set of calibration measurements (where RSD is the relative standard deviation). In this notional example, the CRM LGC1110 (*p,p'*-DDE, with a certified purity of 99.6%) has been used to prepare a series of calibration solutions, which have been measured by an instrumental technique such as gas chromatography with either an electron capture detector or a mass selective detector.

The first step is to plot a graph of the data and carry out a **visual inspection** of the calibration line obtained. As discussed in Section 3.11.1, it is important to plot the data and examine it visually to confirm that it does exhibit linear behaviour. Simply relying on statistical calculations of the correlation coeffi-

Table 4.8   *Calibration data for p,p'-DDE*

Solution	Concentration µg mL^{-1}	Instrument signal						
		1	*2*	*3*	*4*	*mean*	*s.d.*	*RSD*
1	0.2	1060	1130	1095	1135	1105	35	0.032
2	0.4	1960	1835	1910	1935	1910	54	0.028
3	0.6	3340	3365	3240	3135	3270	105	0.032
4	0.8	4250	3980	4110	4020	4090	120	0.029
5	1	5465	5220	5380	5550	5404	141	0.026
6	1.2	6620	6710	6350	6340	6505	188	0.029
7	1.4	7250	7480	7150	7600	7370	206	0.028

cient to check for linearity may lead to erroneous conclusions being drawn. A graph of the data in Table 4.8 is shown in Figure 4.7.

The graph shows that the instrument signal varies linearly with the analyte quantity (*p,p'*-DDE concentration) over the entire range studied. The graph could be used as it is to calculate the *p,p'*-DDE concentration in a sample, by manually interpolating the concentration value using the measured instrument signal and the manually drawn calibration line (Figure 4.8).

However, a more rigorous approach is to calculate the equation of the best straight line using the statistical technique of linear regression (least squares method). This approach also allows the uncertainty of the calibration line to be calculated and consequently the uncertainty in any interpolated concentration value for a sample being analysed.

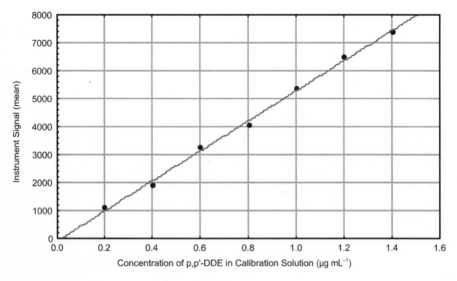

Figure 4.7   *Calibration graph for p,p'-DDE*

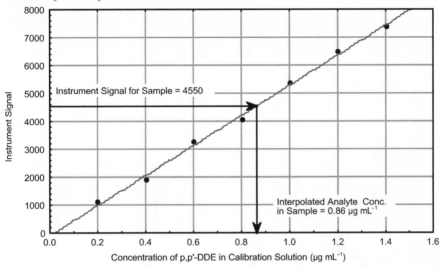

Figure 4.8 *Interpolation using a calibration graph*

4.8 Using Linear Regression to Calculate the Calibration Line

The least squares method of linear regression[4,6-8] is used to fit the best straight line to a set of calibration data such as that shown in Table 4.8. (Chapter 3, Section 3.11 describes the statistical principles underlying the method of linear regression).

The best straight line is that which minimises the sum of the squares of the residual distances of the individual calibration points (measured on the y-axis, *i.e.* instrument signal axis) from the calculated best straight line. This is referred to as the regression of y on x and Figure 4.9 illustrates the residual distances of the calibration points from the calibration line.

The important parameters of the regression line are:

- the slope of the line and its uncertainty
- the intercept of the line and its uncertainty

These parameters are readily calculated using an appropriate software package and a personal computer. For background information, the formulae for the calculations of the uncertainty estimates (as standard deviations) of the slope and intercept of the line are given below. In Eqs. 4.9 and 4.10, s_m is the standard deviation of the slope and s_c is the standard deviation of the intercept respectively.

$$s_m = \frac{rsd}{\sqrt{\sum_i (x_i - \bar{x})^2}}$$

(Eq. 4.9)

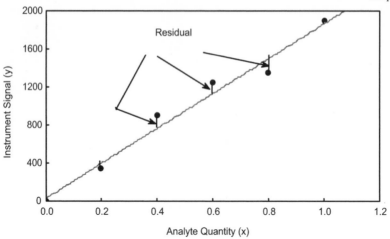

Figure 4.9 *Residuals of calibration points for the regression of y on x*

$$s_a = rsd \sqrt{\left\{ \frac{\sum_i x_i^2}{n \sum_i (x_i - \bar{x})^2} \right\}} \qquad \text{(Eq. 4.10)}$$

$$rsd = \sqrt{\frac{\sum_i (y_i - \hat{y}_i)^2}{n - 2}} = s_y \sqrt{\frac{(n-1)}{(n-2)} (1 - r^2)} \qquad \text{(Eq. 4.11)}$$

The parameter *rsd* is the residual standard deviation (also known as the residual standard error and the standard error of the estimate). It is a measure of the deviation of the data from the regression line (*i.e.* the magnitude of the residuals [see Figure 4.9]). The symbols have the following meaning:

s_m = standard deviation of the slope of the regression line
s_c = standard deviation of the intercept of the regression line
s_y = standard deviation of the measured instrument signals (*y* values)
x_i = the value of *x* for the ith calibration point
$\bar{x}$ = the mean value of *x*
y_i = the measured value of *y* for the ith calibration point
$\hat{y}_i$ = the value of *y* on the regression line for the ith calibration point
r = the correlation coefficient of the regression line
n = number of points in the regression line

Because of the uncertainty in the values for the slope and intercept, there is a corresponding uncertainty in the best straight line that is fitted to the data. The 95% confidence limits for a regression line take the general form shown in Figure 4.10.

It is seen that the uncertainty of a linear calibration line is at a minimum at the mid-point (the point $\bar{x},\bar{y}$) of the line. Therefore, it is recommended that the instrument signal for the sample being analysed should ideally fall near to the

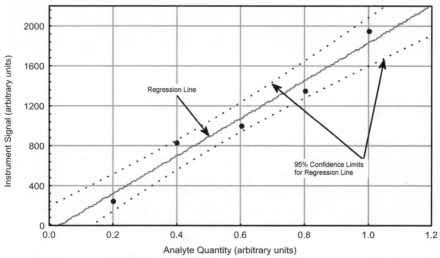

Figure 4.10 *95% confidence limits for a regression line*

mid-point of the calibration line, as discussed in Section 4.7.6. This will serve to minimise the contribution the instrument calibration makes to the overall uncertainty of an analytical result on a routine test sample.

The formal calculation of the uncertainty in an analyte quantity value interpolated from a regression line using a measured instrument signal (Figure 4.7) is a complex problem. The formula that is generally used is:

$$s_X = \left\{\frac{rsd}{m}\right\}\sqrt{\frac{1}{N} + \frac{1}{n} + \frac{(\bar{y}_0 - \bar{y})^2}{m^2(n-1)s_{(x)}^2}} \qquad \text{(Eq. 4.12)}$$

where the symbols have the following meanings:

s_X = standard deviation of the interpolated analyte quantity value for the sample being analysed
m = slope of the regression line
N = number of replicate measurements made on the sample being analysed
n = number of points in the regression line
$\bar{y}_0$ = mean value of the instrument signal for L replicate measurements of the sample being analysed
$\bar{y}$ = mean value of the instrument signal for the n calibration points (*i.e.* the y value at the mid-point of the regression line)
$s_{(x)}$ = standard deviation of the x data (analyte quantity values) for the n points of regression line

Examination of Eq. 4.12 leads to a number of important conclusions:

- The instrument signal response obtained from the sample being analysed should be near to the centre of the calibration line, so that the value

$(\bar{y}_0 - \bar{y})$ is minimised. This will help to minimise the uncertainty in the interpolated value of X (the analyte quantity in the sample analysed).

- Increasing the number of replicate measurements (N), made on a particular sample or sample extract being analysed, will also reduce the uncertainty. However, the law of diminishing returns applies to this approach, so that the number of replicates should typically be in the range 2–5.

- The number of calibration points (n) used to construct the regression line may also be increased as a means of reducing the uncertainty of the estimate of analyte quantity in the sample being analysed. However, unless the preparation of the calibration standards is easy and cheap, there is little benefit to be obtained by using more than 10 calibration points. As discussed in Section 4.7.4, the calibration standards used should, ideally, be prepared independent of each other, in order to take account of all likely sources of uncertainty in their preparation. For example, if a stock calibration solution is prepared and aliquots are then diluted to prepare the other calibration standards, any uncertainties in the preparation of the stock solution will not be taken account of in the computation of the regression line.

These points are illustrated in the examples given below, which are based on the calibration data shown in Table 4.8 and Figure 4.7. The equation of the regression line is:

$$\text{Instrument signal} = 5378.4 \times \text{Analyte quantity} - 66.4 \quad (\text{Eq. 4.13})$$

The mid-point of the regression line $(\bar{x},\bar{y})$ is 0.8, 4236.3.

The tables show the analyte quantity values calculated from Eq. 4.13 for the particular instrument signal values tabulated. Also shown is the standard deviation of the analyte quantity value, calculated according to Eq. 4.12.

It is seen that the absolute error (s_X) is at a minimum when the instrument signal measured for the sample is near to the midpoint of the regression line. (In this example the value of $N = 1$ and $n = 7$ were used in Eq. 4.12 to calculate s_X).

Table 4.10 shows the effect of replicate measurements from which it is seen that about four replicate measurements are optimum, taking account of the improvement in uncertainty and the time required to carry out the measurements. However, if a higher uncertainty is acceptable, single or duplicate

Table 4.9 *Effect of distance from the midpoint of the regression line*

Measured instrument signal ($\bar{y}_0$)	Midpoint of regression line ($\bar{y}$)	Calculated analyte quantity (X) ($\mu g\ mL^{-1}$)	Standard deviation of analyte quantity (s_X)	s_X as a % of X
4550	4236	0.8583	0.0287	3.35
7050	4236	1.3232	0.0316	2.39
1500	4236	0.2912	0.0314	10.80

Table 4.10 *Effect of replicate instrumental measurements*

Measured instrument signal ($\bar{y}_0$)	Number of replicate instrumental measurements (N)	Calculated analyte quantity (X) ($\mu g\ mL^{-1}$)	Standard deviation of analyte quantity (s_X)
4550	1	0.8583	0.0287
4550	2	0.8583	0.0216
4550	4	0.8583	0.0169
4550	8	0.8583	0.0140

instrument readings may be taken, with savings of cost and time (in this example the value of $n = 7$ was used in Eq. 4.12 to calculate s_X).

4.9 Combining Calibration Uncertainty with Analytical Uncertainty

It will be recalled from Sections 4.3.2 and 4.4.2 that there are additional sources of uncertainty in an analytical determination that are **not** taken account of in the calculation of s_X as described in the preceding Section 4.8. These are:

- the uncertainty due to the CRM itself (*e.g.* the uncertainty of the certified property value), $u(CRM)$
- the uncertainty associated with any preparation of the CRM prior to its introduction to the instrument, $u(CRM\ prep.)$
- the uncertainty associated with any preparation of the sample prior to its introduction to the instrument, $u(sample\ prep.)$

An estimate of the overall standard uncertainty of an analytical result is therefore obtained as follows:

$$u(overall) = \sqrt{\{u(CRM)^2 + u(CRM\ prep.)^2 + s_X^2 + u(sample\ prep.)^2\}}\quad \text{(Eq. 4.14)}$$

An example of the use of this equation to combine uncertainties is given below. The example is based on the determination of *p,p'*-DDE in a sample of animal oil at a nominal level of 1 mg kg^{-1}. It is assumed that a 10 g sample was taken and a 10 mL extract prepared for instrumental measurement of the *p,p'*-DDE concentration by GC–MS or GC–ECD. The instrument calibration data are those given in Table 4.7 and Figure 4.7. The calibration solutions are assumed to have been prepared from the CRM LGC1110 (*p,p'*-DDE with a certified purity of 99.6% and an expanded uncertainty [$k = 2$] of 0.4%). The procedures used to prepare the solutions are assumed to be similar to those described in Section 4.5, Table 4.5. The measured instrument signal for the prepared sample extract is taken to be 4550; this figure is assumed to be the mean of four replicate measurements of the extract. The interpolated analyte quantity (*p,p'*-DDE

Table 4.11

Source of uncertainty	Symbol	Units	Value	u	ru	ru²
p,p'-DDE conc. in sample extract	s_X	μg mL^{-1}	0.8583	0.0169	0.01969	0.00038
Purity of p,p'-DDE CRM	u(RM)	%	99.6	0.2	0.002008	4.03E−06
Conc of p,p'-DDE CRM solution	u(RM prep.)	μg mL^{-1}	1	0.008	0.008	0.000064
Preparation of sample extract	u(sample prep.)				0.1	0.01
p,p'-DDE concentration in the original sample	μg g^{-1}		0.8583	0.087764	0.102253	0.010456

u = standard uncertainty; ru = relative standard uncertainty ($= u$/value); ru^2 = relative standard uncertainty squared

concentration in the extract) is therefore 0.8583 μg mL^{-1}, with a standard uncertainty of 0.0169 μg mL^{-1} (Table 4.10).

Using this information, Table 4.11 may be constructed, which itemises each source of uncertainty as a standard uncertainty and shows how the individual uncertainties may be combined. The uncertainty associated with the preparation of the sample has been estimated from the Horwitz function,[9] which predicts an inter-laboratory coefficient of variation of about 10% for the analysis of trace components present in samples at a concentration of 1 mg kg^{-1}.

The individual standard uncertainties are expressed as relative standard uncertainties prior to combination in the usual manner (addition as the squared values). It is seen that the overwhelming contribution to the uncertainty of the final analytical result is due to the uncertainty associated with the sample preparation procedure. The CRM and the calibration procedure make a negligible contribution to the uncertainty of the final result, which is, of course, the preferred situation. In fact, the coefficient of variation associated with the sample preparation would need to be as low as 2% ($ru = 0.02$), before the calibration step begins to make a significant contribution to the overall uncertainty of the final result (which would then be 0.0251). Section 4.3.2 provides further discussion of the required relationship between the calibration uncertainty and the maximum uncertainty ultimately acceptable in an analytical result.

4.10 Assumptions in the Use of Linear Regression

The use of linear regression to construct a calibration line as described in Section 4.8 makes two main assumptions regarding the calibration data.[4,6,8] These are:

1. The errors in the *x*-values (analyte quantity values) are negligible compared with the errors in the measured *y*-values (instrument signal values)
2. The errors in the *y*-values are constant across the entire range of the calibration points

The validity of assumption 1 may be judged by comparing the uncertainties of the analyte quantity values in the prepared calibration standards with the uncertainties in the corresponding instrument signal values. The former may be calculated using an approach of the type illustrated in Section 4.5. The latter may be calculated from measurements of the repeatability of replicate instrument signal measurements. Using the calibration data for p,p'-DDE as an example, it may be calculated from Table 4.8 that the repeatability coefficient of variation for replicate measurements of a particular calibration solution is typically 3%. This may be compared with the uncertainty of the concentration of a calibration solution, which is typically 0.7% (*e.g.* see Table 4.6 [$ru = 0.007054$]). Thus, the assumption may be regarded as being valid in this instance. Generally speaking, calibration solutions may be prepared gravimetrically or volumetrically with small uncertainties, so that unless the instrument readings are comparably precise, the assumption will usually be valid. Procedures for dealing with the situation where the uncertainties of the analyte quantity values are not negligible have been described by MacTaggart and Farwell.[10]

In contrast, assumption 2 may not always be valid since the standard deviation of a set of replicate instrument measurements may increase in proportion with the magnitude of the analyte quantity in the calibration standard being measured. This effect is seen in Table 4.8, where it is the *relative* standard deviation that is approximately constant over the range of calibration solutions, so that the absolute standard deviations progressively increase from the bottom through to the top solution in the range.

An alternative approach in this situation is to calculate a weighted regression line,[6] in which additional weight is given to those calibration points where the error is smallest. The weight for a given calibration point is calculated using Eq. 4.15, where s_i is the standard deviation for replicate measurements of the calibration point concerned and n is the total number of calibration points:

$$w_i = s_i^{-2} / (\sum_i s_i^{-2}/n) \qquad \text{(Eq. 4.15)}$$

The weights have been scaled so that their sum is equal to the number of points on the graph (n).

The slope and intercept of the weighted calibration line are calculated as follows:

$$\text{slope } (b) = \frac{\sum_i w_i x_i y_i - n\bar{x}\bar{y}}{\sum_i w_i x_i - n\bar{x}^2} \qquad \text{(Eq. 4.16)}$$

$$\text{intercept} = \bar{y} - m\bar{x} \qquad \text{(Eq. 4.17)}$$

x_i = the analyte quantity value for the ith calibration standard
y_i = the measured instrument signal for the ith calibration standard
$\bar{x}$ = the mean of the x_i values
$\bar{y}$ = the mean of the y_i values.

Applying these equations to the calibration data in Table 4.8, the following weighted regression line is obtained:

$$\text{Instrument signal} = 5248.7 \times \text{Analyte quantity} - 1.5 \quad \text{(Eq. 4.18)}$$

The 'mid-point' (or weighted centoid) of the weighted regression line $(\bar{x}, \bar{y})$ is 0.3715, 1948.

Comparison with the unweighted regression line (Eq. 4.13) shows that the slopes of the two lines are similar, but that the intercept for the weighted line is nearer to the origin of the graph. Also, the mid-point of the weighted regression line is closer to the origin than for the unweighted line, reflecting the fact that greater weighting is given to calibration points nearer the origin because of the lower absolute standard deviations of these points.

When the weighted regression line is used to calculate the analyte quantity in a sample from the measured instrument signal for that sample, the standard deviation (s_X) of the analyte quantity value is calculated using the equations below[6]

$$s_X = \frac{rsd}{m} \left\{ \frac{1}{w} + \frac{1}{n} + \frac{(\bar{y}_0 - \bar{y})^2}{m^2 \left(\sum_i w_i y_i^2 - n\bar{x}^2 \right)} \right\} \qquad \text{(Eq. 4.19)}$$

$$rsd = \sqrt{\left\{ \frac{\left(\sum_i w_i y_i^2 - n\bar{y}^2 \right) - b^2 \left(\sum_i w_i x_i^2 - n\bar{x}^2 \right)}{n - 2} \right\}} \qquad \text{(Eq. 4.20)}$$

where w is the weighting appropriate to the measured instrument signal $(\bar{y}_0)$ for the sample being analysed.

Using these equations, the analyte quantity values and the corresponding uncertainties have been calculated for instrument signal values of 4550, 7050 and 1500. The results are presented in Table 4.12.

Comparison with the data in Table 4.9 (which was calculated for the unweighted regression line) shows that using the weighted line has relatively little effect on the analyte quantity value (X), but that the uncertainty (s_X) is significantly larger. The effect is most noticeable for the two higher X values, as

Table 4.12

Measured instrument signal $(\bar{y}_0)$	Mid-point of weighted regression line $(\bar{y})$	Calculated analyte quantity (X) $(\mu g\ mL^{-1})$	Standard deviation of analyte quantity (s_X)	s_X as a % of X
4550	1948	0.8672	0.1027	11.84%
7050	1948	1.3435	0.2431	18.09%
1500	1948	0.2861	0.0146	5.10%

the corresponding $\bar{y}_0$ values are significantly distant from the mid-point of the weighted regression line.

4.11 Abridged Methods of Calibration

Construction of a calibration line using seven or more calibration points every time a sample (or batch of samples) is analysed can be a time-consuming and expensive process. Some alternative, shorter calibration procedures[11] are therefore discussed below. However, such procedures should only be applied after it has been established by use of the full calibration procedure described in Sections 4.7 and 4.8 that the calibration graph is linear over the range of interest.

4.11.1 Single Point Calibration

Where previous experience shows that a linear calibration line which passes through the origin may be expected (calibration model 1, Section 4.6.1), a single calibration standard only may be used. The documented property value (*e.g.* concentration) of the calibration standard is chosen to equal or somewhat exceed the maximum value likely to be encountered in the samples being analysed, whilst also being within the already-established linear range of the instrument. Provided the reagents used to prepare the sample for instrumental measurement do not give a significant instrument signal (*i.e.* provided the sample blank is zero), then the analyte quantity in the sample may be calculated using the equation:

$$\text{Analyte quantity} = \frac{\text{Instrument signal for sample}}{\text{Instrument signal for calibrant}} \times \text{Analyte quantity in calibrant}$$

(Eq. 4.21)

If the sample blank value is significant it must first be subtracted from the measured instrument signal values for the sample and calibrant before Eq. 4.21 is used.

4.11.2 Two Point Calibration

Where it is known that the calibration plot is linear, but does **not** pass through the origin of the calibration graph (calibration model 2, Section 4.6.2), then the abridged calibration procedure requires the measurement of two calibration standards, rather than one.[11] The standards should be chosen so that the higher calibrant slightly exceeds the likely maximum analyte quantity in the samples being analysed, whilst remaining within the linear range of the instrument concerned. The lower standard should be chosen to be slightly lower than the likely minimum analyte quantity in the samples being analysed. As in the single point method, if the reagent blank from the sample preparation procedure gives a significant instrument signal reading this must be subtracted from the instrument signals for the two calibration standards and the sample being analysed. The analyte quantity in the sample (X) is interpolated from the calibration line drawn through the two calibration points, or by use of equation 4.22:

$$X = \frac{x_2(\bar{y}_0 - \bar{y}_1) - x_1(\bar{Y} - \bar{y}_2)}{\bar{y}_2 - \bar{y}_1} \qquad \text{(Eq. 4.22)}$$

where:

$\bar{y}_0$ = mean of the instrument signal values for the sample being analysed
$\bar{y}_1$ = mean of the instrument signal values for calibration standard 1
$\bar{y}_2$ = mean of the instrument signal values for calibration standard 2
x_1 = analyte quantity value for calibration standard 1
x_2 = analyte quantity value for calibration standard 2

4.11.3   Uncertainty of Single Point and Two Point Calibrations

There appear to be no recommended procedures for estimating the uncertainty of a sample analyte quantity value obtained from a single point or a two point calibration. In the absence of such information, one approach to estimating the uncertainty (s_X) of the analyte quantity (X) is as follows:

$$s_X = X\sqrt{\frac{rsd_{sample}^2}{L_{sample}} + \frac{rsd_{calibrant1}^2}{L_{calibrant1}} + \frac{rsd_{calibrant2}^2}{L_{calibrant2}}} \qquad \text{(Eq. 4.23)}$$

where:

rsd_{sample} = relative standard deviation of replicate instrument signal measurements on the sample
$rsd_{calibrant1}$ = relative standard deviation of replicate instrument signal measurements on calibrant1
$rsd_{calibrant2}$ = relative standard deviation of replicate instrument signal measurements on calibrant 2*
L_{sample} = number of replicate instrumental measurements made on the sample
$L_{calibrant}$ = number of replicate instrumental measurements made on the calibrant(s)

However, it should be appreciated that the single point procedure implicitly assumes that the calibration is a perfect straight line (correlation coefficient $r = 1$), passing through two points, namely the origin of the calibration graph and that of the single calibration standard. Likewise, the two point procedure assumes perfect linearity through the two calibration standards. These, of course, are unrealistic assumptions as no calibration graph will be perfectly linear. Furthermore, Eq. 4.23 does not take account of the distance of the sample instrument signal from the calibrant signal (*cf.* Table 4.9). Consequently, the uncertainty estimate given by Eq. 4.23 may well be an under-

---

*For a single point calibration, there is no calibrant 2.

estimate, depending on the scatter of the calibration points about the actual calibration line and difference between the sample and calibrant signals.

It should also be remembered that to obtain the total standard uncertainty of the analyte quantity value in the sample, the value of s_X obtained above must be combined with the the additional sources of uncertainty discussed in Section 4.9.

4.12 Calibration by Bracketing

Where there is some doubt about the linearity of the calibration graph, or the stability of the instrument readings with time, the bracketing technique[4] may be useful. This requires obtaining a preliminary estimate of the analyte quantity in the sample by use of one of the calibration procedures described in Sections 4.7–4.11. Two further calibration standards are then prepared with property values (*e.g.* concentrations) that are slightly smaller and slightly larger respectively than the analyte quantity expected in the sample. The aim is to bracket the sample as closely as possible, so that the measurements are carried out over a very small portion of the calibration line, thereby minimising errors due to any non-linearity in the calibration graph. The analyte quantity (X) in the sample is calculated using Eq. 4.22.

The standard uncertainty (s_X) of the analyte quantity is calculated thus:

$$s_X = \sqrt{\frac{\sum_{k=1}^{K}(y_{1k} - \bar{y}_1)^2 + \sum_{k=1}^{K}(y_{2k} - \bar{y}_2)^2 + \sum_{k=1}^{K}(y_k - \bar{y})^2}{3(K-1)}} \qquad \text{(Eq. 4.24)}$$

where:

y_{1k} = instrument signal for the kth replicate measurement of calibration standard 1

y_{2k} = instrument signal for the kth replicate measurement of calibration standard 2

$\bar{y}_1$ = mean instrument signal for K measurements of calibration standard 1

$\bar{y}_2$ = mean instrument signal for K measurements of calibration standard 2

y_k = instrument signal for the kth replicate measurement of the sample being analysed

$\bar{y}$ = mean instrument signal for K measurements of the sample being analysed

To obtain the total standard uncertainty, the value s_X must be combined with the other relevant sources of uncertainty, as discussed in Section 4.9.

4.13 Matrix Effects and Calibration

All of the foregoing calibration procedures are based on the implicit assumption that the quantitative response of the instrument to the calibration standards is the same as the response to the sample being analysed. However, as discussed in Sections 4.3.1 and 4.4.5, matrix effects may be present, such that the sample matrix leads to either a suppression or enhancement of the sample signal compared to the calibrant signal for the same analyte quantity. Sample matrix effects are particularly prevalent in such techniques as atomic absorption spectroscopy (with both flame and furnace sources), atomic emission spectroscopy (with arc, spark and plasma sources), X-ray fluorescence spectrometry and electrochemical techniques such as anodic stripping voltammetry. If such effects are present they must be taken account of in the calibration procedure, otherwise biased results will be produced, no matter how thoroughly the remainder of the calibration is carried out. A number of possible approaches are discussed below.

4.13.1 Sample Dilution

One of the simplest approaches to matrix effect problems is to dilute the prepared sample extract, using a solution similar to that used to prepare the calibration standards. Such dilutions may be sufficient to eliminate any effects the sample matrix may have on the observed instrument response. The viability of this approach depends on the analyte levels in the sample being sufficiently high so that the necessary dilution does not bring the analyte levels below or too close to the instrument detection limit.

4.13.2 Matrix Matching

In this approach, the composition of the matrix in which the calibration standards are prepared is adjusted to approximate to the matrix of the sample (or that of the prepared sample extract) that is presented to the instrument. For example, if the metal content of a foodstuff is being determined by atomic emission spectroscopy, following digestion of the foodstuff in nitric acid, the calibration standards would be prepared in nitric acid of an appropriate concentration. If sufficient information is available regarding the presence of other species in the sample extract (*e.g.* salts such as NaCl), then these also may be added in appropriate amounts to the calibration standards prior to their introduction to the instrument.

4.13.3 Standard Additions

Where the composition of the sample is not sufficiently known to allow an appropriate matrix match to be obtained for the calibration standards, the technique of standard additions may be used. This involves taking a number of aliquots, of equal volume, of the prepared sample extract and adding to all but one of them accurately known quantities of the calibration standard. The samples thus prepared, after they have all been adjusted to the same volume after addition of an appropriate pure solvent, are then subjected to instrumental measurement in the normal way. The process of standard additions effectively produces a calibration graph using the actual sample being analysed as the calibration matrix. Thus a 'perfect' match is obtained, so that the analyte due to

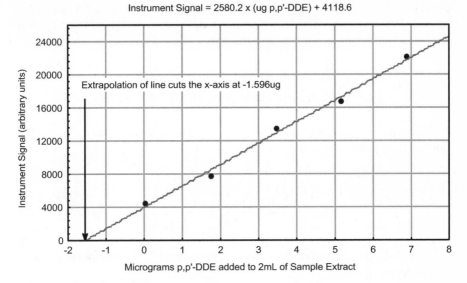

Figure 4.11 *Graph of standard additions*

the calibrant may be expected to behave in an identical way to the analyte due to the sample.

The measured instrument signal is then plotted (*y*-axis) *versus* the known quantity of calibrant added (*x*-axis). A graph of the type shown in Figure 4.11 is obtained. Least squares linear regression is used to fit the best straight line to the data points.

This example refers to the analysis of a sample of animal fat for *p,p′*-DDE content. A 10 g sample was taken and a 10 mL extract was prepared. The extract was divided into five 2 mL portions and accurately measured quantities of the *p,p′*-DDE calibration standard were added to each of the solutions. One of the 2 mL portions was left unspiked. The volumes of all the extracts were then made to 3 mL prior to the instrumental measurements, the results of which are given in Table 4.13.

The amount of *p,p′*-DDE in the unspiked 3 mL extract is determined by extrapolation of the line so that it cuts the *y*-axis at *y* = 0. The value of *x* at this

Table 4.13

µg p,p′-DDE added	*Instrument signal*
0	4550
1.72	7865
3.44	13465
5.16	16841
6.88	22252

point (1.596 μg) gives the amount of *p,p'*-DDE in the original 2 mL unspiked extract. This value may also be obtained by calculation, using the intercept (on the *y*-axis) and slope of the best straight line:

$$\mu g\ p,p'\text{-DDE in 2 mL extract} = \text{intercept/slope}$$

$$\mu g\ p,p'\text{-DDE in 2 mL extract} = 4118.6/2580.2 = 1.596\ \mu g$$

To calculate the standard uncertainty (s_X) of the analyte quantity value, the following equation is used:[6]

$$s_X = \left\{\frac{rsd}{m}\right\}\sqrt{\frac{1}{n} + \frac{\bar{y}^2}{m^2(n-1)s_{(x)}^2}} \qquad (\text{Eq. } 4.25)$$

where:
$s_{(x)}$ = standard deviation of the *x* data (analyte quantity values) for the *n* points of regression line
m = slope of the regression line
n = number of points in the regression line
$\bar{y}$ = mean value of the instrument signal for the *n* calibration points

and *rsd*, the residual standard deviation is calculated as shown in Eq. 4.11.

Using the above equation, a value of 0.270 μg is calculated for the standard uncertainty of the analyte value (1.596 μg), obtained by extrapolation of the standard additions graph. To obtain the total standard uncertainty, the value s_X must be combined with the other relevant sources of uncertainty, as discussed in Section 4.9.

The use of the standard additions procedure requires that the calibration graph as normally prepared is linear and passes through the origin (*i.e.* calibration model 1, Section 4.6.1). When making the additions, it is recommended that the first addition (*i.e.* the smallest addition) should be such that the analyte quantity is increased by about 100%. Subsequent additions are normally the same as the first, so that the pattern shown in Table 4.12 is typical for a standard additions exercise. It is seen from Eq. 4.25 that the standard uncertainty decreases as the number of points in the line (*n*) increases. Thus six points would normally be considered a reasonable optimum.

4.14   Detection of Non-linearity

Whilst visual inspection of a calibration graph is often capable of establishing any departures from linearity (*e.g.* see Figure 4.4), there may be situations where minor deviations from linearity occur that are not easily detected in this way. Under these circumstances it is useful to construct a graph that plots residual values[7] *versus* the observed values for each of the calibration standards used to construct the calibration graph. A residual value is the difference between the

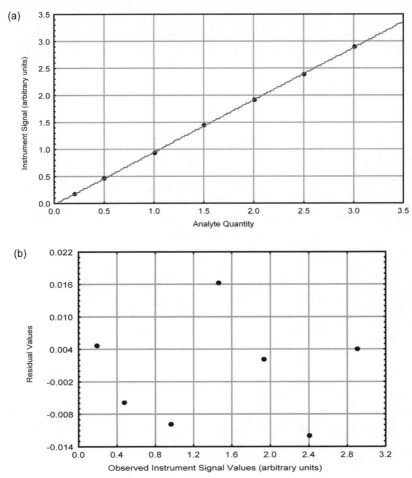

Figure 4.12 (a) *linear calibration graph;* (b) *observed values vs. residuals*

observed instrument signal value for a particular calibration standard and the corresponding theoretical value given by the regression equation (see also Figure 4.9).

Figure 4.12a shows a typical linear calibration graph and Figure 4.12b shows the residual plot for the same data. The random way in which the residual values are distributed is typical of a linear calibration relationship. To avoid possible confusion, the reader should note that the *x*-axes of the two graphs do not represent the same parameter, even though the numerical values are similar.

Figure 4.13a shows a calibration graph that exhibits departure from linearity at higher analyte quantity values, while Figure 4.13b shows the residual plot. The latter shows the pattern that is diagnostic of departures from linearity of the type shown in Figure 4.13a.

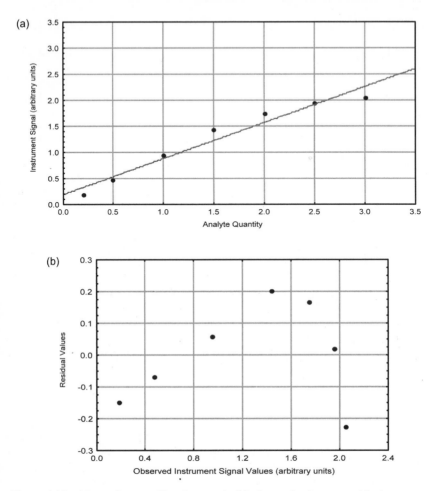

Figure 4.13 (a) *non-linear calibration graph;* (b) *observed values vs. residuals*

Thus, construction of a residual plot may be a useful means of detecting departures from linearity, especially where they are less obvious than shown in Figure 4.13a and not so readily detectable by straightforward visual inspection. An example of this is shown in Figures 4.14a and 4.14b. Whilst the departure from linearity is by no means obvious in Figure 4.14a, the residual plot shows a definite pattern, indicating that the data are non-linear. The graph actually follows a quadratic equation such that the plot of the data is really a curve with a very slight minimum (compared with the notional straight line forced through the data) at its mid-point. The pattern of the residual plot is thus opposite in form to that in Figure 4.13b. Figure 4.14c shows the quadratic curve that gives the best fit to all of the data points.

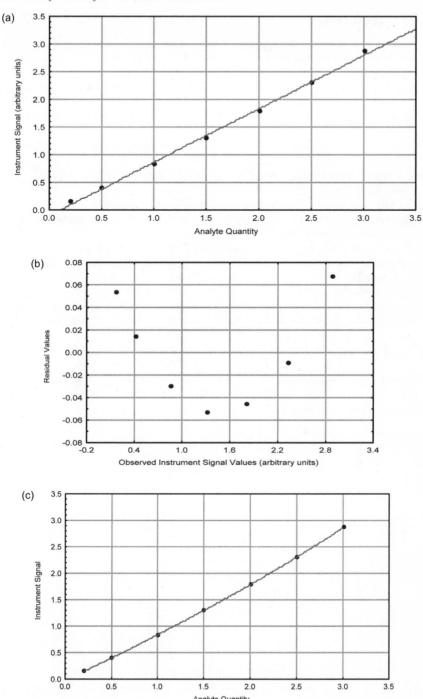

Figure 4.14 (a) *non-linear calibration graph;* (b) *observed values vs. residuals;* (c) *quadratic calibration graph*

4.15 References

1. *Analytical Chemistry: The Approved Text to the FECS Curriculum*, ed. R. Kellner, J.-M. Mermet, M. Otto and H. M. Widmer, Wiley-VCH, 1998 (ISBN 3-527-28881-3). Chapter 12.2, Calibration.
2. *Certification of Reference Materials – General and Statistical Principles*, ISO Guide 35, 1989.
3. *Quantifying Uncertainty in Analytical Measurement*, EURACHEM, 1995 (ISBN 0-948926-08-2).
4. *Linear Calibration Using Reference Materials*, International Standard BS ISO 11095, 1996.
5. *Calibration in Analytical Chemistry and Use of Certified Reference Materials*, ISO Guide 32, 1997.
6. *Statistics for Analytical Chemistry* (3rd edition), J. C. Miller and J. N. Miller, Ellis Horwood, 1993 (ISBN 0-13-030990-7). Chapter 5, Errors in Instrumental Analysis; Regression and Correlation.
7. Regression and Calibration, S. Burke, *VAM Bulletin*, No.18, pp. 18–21, 1998.
8. Analytical Use of Linear Regression. Part I: Regression Procedures for Calibration and Quantitation, D. L. MacTaggart and S. O. Farwell. *J. AOAC Internat.*, 1992, **75**, No. 4, 594–607.
9. Quality Assurance in the Analysis of Foods for Trace Constituents, W. Horwitz, L. R. Kamps and K. W. Boyer, *J. Assoc. Off. Anal. Chem.*, 1980, **63**, No. 6, 1344–1354.
10. Analytical Use of Linear Regression. Part II: Statistical Error in Both Variables, D. L. MacTaggart and S. O. Farwell, *J. AOAC Internat.*, 1992, **75**, No. 4, 608–614.
11. *Samples and Standards, Analytical Chemistry by Open Learning*, B. W. Woodget and D. Cooper, John Wiley and Sons, 1987 (ISBN 0-471-91289-1). Chapter 6, Analytical Standards and Calibration Curves.

CHAPTER 5

Use of CRMs for Assessing the Accuracy of Analytical Data

5.1 The Importance of CRMs in Routine Analysis

Even when a laboratory is using a standardised analytical method, with documented performance characteristics for trueness and precision, established according to recognised method validation protocols,[1,2] there is still a need for a practical demonstration that the accuracy of its results are acceptable and fit-for-purpose. The need is even greater if the analytical method being used has not been the subject of a full validation exercise. There are various ways of demonstrating that routine results are of the required accuracy, including participation in proficiency testing schemes, the analysis of 'spiked' samples, the application of two or more methods and *ad hoc* sample exchanges between two or more laboratories.

In addition to these approaches, the analysis of an appropriate matrix CRM offers a particularly effective means of assessing and demonstrating the quality of analytical data. A matrix CRM resembles the routine samples being tested but, in addition, contains known analytes with documented property values (*e.g.* concentration values). By applying an analytical method to the matrix CRM and comparing the result obtained with the documented concentration value for the analyte(s) of interest, any bias in the result can be detected and, if necessary, appropriate corrective action can be taken. It is important to note that when used in this way, matrix CRMs enable any bias due to a laboratory's execution of a particular method to be identified. There is an implicit assumption that the method itself is capable of producing unbiased data. Indeed, it would not be expected that a laboratory would use a method for routine work if its fitness-for-purpose had not been established by a prior method validation exercise of an appropriate type. If method validation has not been carried out, one approach is to use a matrix CRM as described in Chapter 6.

5.2   The Selection of Matrix CRMs

When a matrix CRM is selected for the purpose of assessing the accuracy of a laboratory's routine analytical results a number of factors should be borne in mind[3] and these are discussed below.

5.2.1   Documentation for the CRM

The CRM should be accompanied by appropriate documentation, such as a certificate, that provides the following essential information:

1. Name and address of the producer of the material
2. Description/name of the material
3. Physical/chemical form of the material
4. Sample number/batch number/certificate number
5. Date of production/issue
6. Shelf life/expiry date
7. Property values
8. Uncertainty of property values
9. Procedures used to characterise the material, *i.e.* to determine its property values

5.2.2   Physical/Chemical Form

The matrix of the CRM should resemble as closely as possible the matrix of the routine samples being analysed. In this way, the CRM replicates any sample characteristics, which may affect the analysis (*e.g.* speciation, extraction efficiency, completeness of matrix dissolution and interferences from matrix components other than the analytes of interest). Under these circumstances a result for the CRM that adequately agrees with the documented value gives good grounds for concluding that the results on the routine samples are fit-for-purpose.

Ideally, the matrix of the CRM should be identical to that of the routine samples, but this will rarely be possible. For example, foodstuff CRMs may have been produced in a freeze-dried form (to enhance their stability) which is markedly different from the form of the fresh foodstuffs that may be undergoing analysis. Likewise, environmental CRMs such as soils and sediments may have been dried to a particular moisture content and ground to a specific particle size distribution that may well be different from that which a laboratory would normally use in the preparation of its routine samples. In yet other types of matrix CRMs such as drinking waters, blood serum, *etc.*, chemical stabilisers may have been added which would not normally be present in the routine samples. Users should therefore be aware that differences of the general type outlined above may affect the analytical procedure to the extent that the results obtained on the CRM may not necessarily be a good guide to the accuracy of the routine results. It is the responsibility of the user to ensure that the matrix CRM selected is appropriate for its intended use.

5.2.3 Documented Property Values

The documented property values of the matrix CRM should be similar to values being sought in the routine samples. Thus, if a soil sample is being analysed for the presence of polychlorobiphenyls (PCBs) at concentrations around 0.1 mg kg^{-1}, the CRM should contain PCBs at about this level. There would be relatively little benefit in selecting a contaminated soil CRM that contains PCBs at a level of 100 mg kg^{-1}, since the analytical problems presented by the two situations could be quite different.

5.2.4 Uncertainty of the Documented Property Values

The intention of using a matrix CRM is to assess any bias in a laboratory's execution of a particular method. This is discussed in detail in Sections 5.3 and 5.4, but essentially it entails comparing the analysed result for the CRM with its documented value; the difference between the two represents a possible bias. However, the documented property value will have an associated uncertainty, which places a constraint on the minimum bias that can be detected by use of *that particular certified reference material.* The rule is that the minimum bias detectable is equal to twice the standard uncertainty (u) of the documented property value quoted on the documentation accompanying the matrix CRM. It should be appreciated, however, that the minimum detectable bias is also dependent on the precision of the measurement method that is being evaluated for bias. Where the method uncertainty is larger than the CRM uncertainty (which will often be the case), the minimum bias that is detectable will be greater than $2u$. The relative importance of the CRMs uncertainty and the method uncertainty is discussed further in Section 5.4.2.

5.2.5 Storage Requirements and Shelf Life/Expiry Date

When selecting and using a matrix CRM attention needs to be paid to the physical and chemical stability of the material (see also Section 4.4.6). For a certified reference material, information should be provided on the certificate and it is important that any special storage requirements are observed. Thus, it may be necessary to store the material in the dark, in a desiccator, in the absence of air, at sub-ambient temperatures (*e.g.* 4 °C, −20 °C, −70 °C), *etc.* Users should ensure that they have the necessary facilities for the proper storage of the CRM, otherwise the documented property values may become invalid.

Another important consideration during storage (and use) of a CRM is the need to scrupulously avoid contamination of the material. If this occurs, the documented property values will again be rendered invalid. Such contamination can take the form of moisture pick-up, chemical contamination by storage adjacent to incompatible chemicals or careless dispensing of the material using contaminated glassware, spatulas, *etc.*

It is also important to observe any expiry date quoted by the producer of the CRM, as the documented property values cannot be guaranteed to be valid after this date.

5.2.6   Quantity Taken for Analysis

It is important that the user uses an appropriate quantity of the matrix CRM, that is one that complies with any stipulation given on the documentation that accompanies the material. Thus, the producer of a matrix CRM will often state that the documented property values only apply when a specified minimum quantity of the material is analysed (a quantity larger than the specified minimum may be taken for analysis). This requirement arises because an important part of the characterisation of a matrix CRM is an evaluation of its homogeneity, and the apparent homogeneity of a material can vary with the sample size taken for analysis. As the documented property values include a contribution for any inhomogeneity in the material, they are only applicable to sample quantities equal to, or greater than, the quantities used by the producer to evaluate the homogeneity of the material.

5.2.7   The Number of Different Matrix CRMs to be Used

If the routine samples being analysed are likely to contain the analytes of interest across a range of concentrations, such that the range exceeds a factor of perhaps 5 or more then, ideally, two or three matrix CRMs should be used. The CRMs should have a range of documented property values that matches that expected in the samples. This is because any bias may be constant, level related, or a combination of these effects, over the range of analyte levels of interest.[5] A problem likely to be encountered, however, is lack of availability of suitable CRMs. In those situations where only one CRM is available, the analyst could consider extending the range by spiking a suitable matrix to approximate to the analyte levels not covered by the CRM. Whilst not ideal, this approach is preferable to relying on just one matrix CRM to check for bias over a wide range of analyte levels.

5.3   Principles Involved in Assessing the Accuracy of Routine Data

5.3.1   What Is Accuracy?

The accuracy of an analytical result is characterised by the closeness of the agreement between the result and the true value (see Chapter 3, Section 2). The true value is the actual concentration of an analyte in a sample matrix and it is the value that would be obtained by a perfect measurement.[6] True values, by their nature, are indeterminate, but the documented property values of CRMs are considered to be valid estimates of the true values, within the stated uncertainties.

Broadly speaking, two factors may be regarded as contributing to the accuracy of a result:[3]

- the trueness (*i.e.* lack of bias) of the analytical method, as executed (systematic error)
- the precision of the analytical method, as executed (random error)

The prime use of a matrix CRM is to assess a result for the presence of bias (systematic error). The bias (Δ) of an analytical result ($\bar{x}$) is given by:

$$\Delta = \bar{x} - \mu \qquad \text{(Eq. 5.1)}$$

where μ is the 'true' value (*i.e.* the documented property value of the CRM).

Because of the presence of random error, the observed value of Δ is likely to be some value other than 0, even when there is no bias present in the result $\bar{x}$. The larger the random error, the larger the value of Δ can be without it being significantly different from 0. That is, the larger the random error, the more difficult it is to detect relatively low levels of bias.

Consequently, as part of the assessment of bias it is necessary also to have some knowledge of the precision (random error) of the result. Approaches to making the necessary initial assessment of precision are discussed below.

5.3.2 Procedures for the Assessment of the Precision of a Measured Result on a Matrix CRM

The precision of a measurement result may be regarded as arising from two contributing factors:

- the within-laboratory standard deviation of the method (s_w)
- the between-laboratory standard deviation of the method (s_b)

Both components of precision should be taken into account when the bias of an analytical result is assessed.[3]

The within-laboratory standard deviation (s_w), which is equivalent to the repeatability standard deviation (s_r), may be estimated from the replicate measurements made on the CRM as part of the bias assessment. To obtain a reliable estimate of s_w, it is recommended that at least seven replicate measurements are carried out (see Chapter 3, Section 3), under repeatability conditions. That is, the measurements are carried out by the same analyst, using the same method and equipment, in the same laboratory environment over the shortest practical time-scale.

The between-laboratory component of the precision is less easy to estimate. The following approaches are suggested, in order of preference.

1. The long-term standard deviation (intermediate precision) of a set of replicate measurements (at least seven and preferably up to 20) made on a CRM over an extended period of time (> 3 months), in one laboratory often provides an acceptable approximation to the between-laboratory reproducibility standard deviation.[3]
2. If the analytical method being used has been the subject of an inter-laboratory validation exercise conducted according to recognised protocols,[1,2] the values for the repeatability standard deviation (s_r) and reproducibility standard deviation (s_R) quoted in the standard document for the method may be used to estimate s_b, thus:

$$s_b = \sqrt{(s_R^2 - s_r^2)} \qquad \text{(Eq. 5.2)}$$

3. Where the CRM has been characterised by means of an inter-laboratory exercise, relevant information on the between-laboratory standard deviation may be given on the documentation accompanying the material.[3] If the analytical method being used by the laboratory wishing to assess its individual bias is similar to the method(s) used in the characterisation of the CRM, any between-laboratory standard deviation quoted in the CRM documentation could be adopted as an estimate of s_b.
4. The inter-laboratory standard deviation predicted by the Horwitz function[7] provides a further means of estimating s_b.
5. In the absence of any other information, an estimate of s_b may be obtained from the measured s_w value. Thus:

$$s_b \approx 2 \times s_w \qquad \text{(Eq. 5.3)}$$

The precision (σ) of a laboratory's measured result on a matrix CRM is calculated by combining the two components in the following manner:[3]

$$\sigma = \sqrt{s_b^2 + \frac{s_w^2}{n}} \qquad \text{(Eq. 5.4)}$$

where n is the number of replicate measurements made on the CRM.

Generally speaking, s_w may be expected to be smaller than s_b (typically by a factor of about 2 – see Eq. 5.3).[7] Combined with the fact that n will normally be at least 7, this means that s_b is usually the major contributor to σ, to the extent that s_w can usually be ignored.

5.3.3 The Rationale for Combining s_b and s_w

At first sight, it may seem odd that the estimated precision of a single laboratory's result, σ, includes a contribution from the between-laboratory variation, s_b. Should not the within-laboratory variation, s_w, measured by replicate determinations on the matrix CRM, be an adequate measure of the expected random dispersion of measured results about the true value?

In answering this question, it should be appreciated that s_w measures the random dispersion of a laboratory's replicate results about the mean of those results. The mean itself is randomly distributed about the true value for the CRM with a dispersion that is characterised by s_b. Thus, s_w and s_b combined (as in Eq. 5.4) is used to describe the overall dispersion of results about the true value.

The parameter s_b measures those sources of random error that cannot be evaluated by replicate measurements in a single laboratory, but which still contribute to the dispersion of an individual laboratory's result about the true value. An example of such a source of random error would be the final volume of a prepared sample extract, before it is introduced to an instrument for measurement, where no account has been taken of the ambient temperature in the laboratory (a common occurrence). For measurements made under repeatability conditions, variations in ambient temperature may be insignificant and therefore not included in s_w. However, the same measurements made in different laboratories (or indeed in a single laboratory over a period of time) could be the subject of additional random error due to variations in ambient temperature. The effect of such variations would be included in s_b.

It may also be helpful to realise that when a laboratory analyses a matrix CRM it is effectively taking part in an 'inter-laboratory comparison', since the documented value for the CRM will have been established by measurements carried out in one or more other laboratories. Under these circumstances it is obviously appropriate that the between-laboratory component of precision should be taken into account when an individual laboratory compares its result to the documented value for the CRM.

A further useful analogy is found in the field of proficiency testing (PT). Often a laboratory's result on a PT sample is compared to the true value (usually the consensus mean of the participants' results) by means of a z-score. This requires adopting a value for the target standard deviation for the measurement and the expected inter-laboratory standard deviation (s_b) is often used for this purpose.[8]

5.3.4 Is the Within-laboratory Precision Acceptable?

If information is available on the within-laboratory standard deviation expected for the method (*e.g.* the repeatability standard deviation (s_r) quoted in the standard document for a method validated in an inter-laboratory exercise according to recognised protocols), then the Chi-squared test may be carried out (see Chapter 3, Section 5). This test will establish whether the value of s_w obtained is acceptable, *i.e.* whether the laboratory's execution of the method is of sufficient precision. However, even when s_w is found to be significantly larger than s_r, if the measured value $s_w^2/\sqrt{n}$ is small compared to s_b^2, there is little or no benefit to be gained in repeating the set of measurements on the CRM with the aim of achieving a lower value of s_w.

5.3.5 How Is a Value for the Bias Calculated?

To answer this key question we first recall from Eq. 5.1 that the apparent bias (Δ) of a result is the difference between the observed result ($\bar{x}$, the mean of n measurements) and the true value (μ).

$$\Delta = \bar{x} - \mu \qquad \text{(Eq. 5.1)}$$

Because of the presence of random error (σ) in the result $\bar{x}$, it is likely that a value for the apparent bias of other than 0 will be obtained, even if the laboratory's execution of the method is unbiased. Consequently, a statistical test is required to assess whether the observed bias (Δ) is significantly different from 0. If it is not, it may be concluded that there is no evidence for bias in the laboratory's application of the method. A suitable test is given in Eq. 5.3, which summarises the fact that if the observed value for the apparent bias Δ falls within the limits $\pm 2\sigma$, there is no evidence for the presence of bias[3] (see also Section 3.6).

$$-2\sigma < \bar{x} - \mu < 2\sigma \qquad \text{(Eq. 5.5)}$$

Careful attention should be paid to the exact form of the conclusion that there is no evidence for the presence of bias. This is not the same as concluding that the method is unbiased; there could be bias present that is beyond the power of the test to detect. Also, with limits of $\pm 2\sigma$ having been adopted in Eq. 5.5, the confidence level attaching to the conclusion is approximately 95%. If $\pm 3\sigma$ had been adopted as the limits, the level of confidence would be 99.7%.

It is informative to note, in passing, that Eq. 5.5 is essentially similar to the z-score that is used in PT schemes to assess the acceptability of participants results. The z-score is given by:[8]

$$z = (\bar{x} - \hat{X})/\sigma \qquad \text{(Eq. 5.6)}$$

where $\hat{X}$ is the true value for the PT material, σ is the target standard deviation for the measurement, both values being established by the scheme organiser, and $\bar{x}$ is the result on the PT material reported by the participant laboratory.

The value for σ is usually based on the expected inter-laboratory standard deviation for the measurement and a z-score of less than 2 is usually taken as indicating that the result $\bar{x}$ is of acceptable accuracy.

It is important that the value used for σ in Eq. 5.5 is a reliable estimate of the precision of the measurement. Thus, of the procedures listed in Section 5.3.2, if procedure 1 is used to estimate s_b, at least seven replicate measurements are required and preferably up to 20, if time and resources permit (see Section 3.6). The estimate of s_w should be based on at least seven replicate determinations, carried out under repeatability conditions. Because the influence of s_w on σ is relatively minor, seven replicate determinations should usually suffice, although additional measurements do no harm if resources permit.

However, if the method used has been the subject of a previous study by the laboratory, such that the value of the within-laboratory repeatability standard deviation is already well established for the matrix of interest, then this value may be used for s_w in Eq. 5.4. The number of determinations on the matrix CRM can be a value less than seven, although duplicates are recommended as a minimum. A single determination could be used, but the laboratory would need to be confident that the measurement was under statistical control, so that the within-laboratory standard deviation of the result really is a good estimate of s_w. The value of n in Eq. 5.4 should, of course, be chosen to reflect the number of replicate measurements actually made on the CRM.

Procedure 2 should provide a well-founded estimate of s_b, although a check should be made to ensure that the measured value of s_w (based on at least seven replicates) is not significantly different from the repeatability standard deviation (s_r) quoted for the method. The Chi-squared test (see Section 5.3.4) should be used for this purpose.

The remaining procedures in Section 5.3.2 should be used with caution and the analyst should endeavour to confirm the estimate of s_b as soon as resources permit using either procedure 1 or 2.

Worked example
The CRM LGC1004 is a water matrix containing triazine herbicides at certified concentrations. For the herbicide simazine, the certified concentration is 26.7 μg kg^{-1}, with an expanded uncertainty ($k = 2$) of 2.0 μg kg^{-1}. Six replicate analyses were carried out on this material, with the results shown in Table 5.1. The value adopted for s_b is 5.2 μg kg^{-1}, based on the measured long-term standard deviation of the procedure.

The calculated value for σ is given by:

$$\sigma = \sqrt{5.2^2 + \frac{2.5^2}{6}} = 5.3 \ \mu\text{g kg}^{-1}$$

The calculated value for the apparent bias is given by: $25.3 - 26.7 = -1.4 \ \mu\text{g kg}^{-1}$.

Table 5.1

Replicate 1	29.4 μg kg^{-1}
Replicate 2	24.9 μg kg^{-1}
Replicate 3	26.4 μg kg^{-1}
Replicate 4	25.7 μg kg^{-1}
Replicate 5	22.0 μg kg^{-1}
Replicate 6	23.5 μg kg^{-1}
n	6
mean	25.3 μg kg^{-1}
s_w	2.5 μg kg^{-1}

It is seen that the measured value for the apparent bias meets the condition of Eq. 5.5, *viz.*

$$-10.6 < 1.4 < 10.6$$

We therefore conclude that there is no evidence for the presence of bias in the laboratory's execution of the method. It should be noted that this is *not* the same as concluding that the method and its execution are unbiased. There could be bias present that is beyond the power of the test to detect. Procedures for calculating the undetected bias that could be present are discussed in Section 5.4.4. We should also recall that the validity of the test depends entirely on the validity of the values adopted for s_b and s_w. If these values are wrong the value for σ will also be wrong and the test will lead to false conclusions being drawn.

5.4 Issues Involved in Assessing the Accuracy of Routine Data

A number of supplementary issues and questions can arise in assessing any possible bias in a laboratory's execution of a particular measurement method. Some of these issues are discussed below.

5.4.1 Should a Result Obtained on a Matrix CRM Lie Within the Uncertainty Range of the Documented Reference Value?

It is sometimes concluded that if a result obtained by a laboratory on a matrix CRM falls outside the uncertainty range of the documented property value, then the result must be biased. However, this conclusion may be wrong since it takes no account of the random uncertainty (σ) in the result as obtained by the laboratory. The uncertainty range of the reference material only addresses the uncertainties arising in the characterisation of the CRM. It does *not* deal with the uncertainty associated with measurements made on the CRM in the user's laboratory.

Worked example
The CRM LGC7103 is a sweet digestive biscuit matrix with certified values for the concentrations of various nutritional components. The certified value for the nitrogen (protein) content is 1.08 g $(100 \text{ g})^{-1}$, with an expanded uncertainty ($k = 2$) of 0.01 g $(100 \text{ g})^{-1}$. Six replicate measurements were carried out on the material with the following results shown in Table 5.2. The value adopted for s_b is 0.015 g $(100 \text{ g})^{-1}$, based on the long-term standard deviation of the method.

The calculated value for σ is given by:

$$\sigma = \sqrt{0.015^2 + \frac{0.004^2}{6}} = 0.015 \text{ g } (100 \text{ g})^{-1}$$

The calculated value of the apparent bias is: $1.096 - 1.08 = +0.016$ g $(100 \text{ g})^{-1}$.

Table 5.2

Replicate 1	$1.091 \text{ g } (100 \text{ g})^{-1}$
Replicate 2	$1.094 \text{ g } (100 \text{ g})^{-1}$
Replicate 3	$1.097 \text{ g } (100 \text{ g})^{-1}$
Replicate 4	$1.099 \text{ g } (100 \text{ g})^{-1}$
Replicate 5	$1.100 \text{ g } (100 \text{ g})^{-1}$
Replicate 6	$1.095 \text{ g } (100 \text{ g})^{-1}$
n	6
mean	$1.096 \text{ g } (100 \text{ g})^{-1}$
s_w	$0.004 \text{ g } (100 \text{ g})^{-1}$

It is seen that the apparent bias meets the requirements of Eq. 5.3, *viz.*

$$-0.03 < 0.016 < 0.03$$

We therefore conclude that there is no evidence for the presence of bias in the method, as executed by the laboratory.

However, the measured value of 1.096 lies *outside* the uncertainty range for the reference material, (1.08 ± 0.01), confirming that it is *not* a requirement that all results *must* fall within this range if it is to be concluded that there is no evidence for bias.

5.4.2 When Is the Uncertainty of the CRM Important?

Although the discussion in the previous section showed that the uncertainty range of the CRM does not define the acceptable limits for results obtained by users of the material, the uncertainty of the reference value is used in other ways. In particular, it is important to establish the significance of the uncertainty of the documented reference value in relation to the estimated uncertainty (σ) of the result obtained on the CRM by the user laboratory. In those situations where the uncertainty of the reference value is likely to be significant compared with the estimated uncertainty (σ) of the result obtained by a user laboratory, the uncertainty of the CRM must be taken into account when the result is assessed for the presence of bias.

Rather than simply using the value σ (as defined by Eq. 5.4) to calculate the acceptable range of results (from Eq. 5.5), an additional term needs to be included in the calculation of σ. The additional term represents the standard uncertainty of the documented value of the CRM. The standard uncertainty of the CRM is calculated from the documented uncertainty, which is normally an expanded uncertainty (U) with a stated coverage factor (k), (see Section 3.7). If a coverage factor of 2 is used, the true value is thought to lie in the range given by the expression (documented value $\pm$ expanded uncertainty) with a 95% level of confidence. If the coverage factor used is 3, the level of confidence is 99.7%. The standard uncertainty (u) of the CRM is calculated as:

$$u_{CRM} = U/k$$

The value of σ used to calculate the acceptable range of results is then given by:

$$\sigma = \sqrt{s_b^2 + \frac{s_w^2}{n} + u_{CRM}^2} \qquad \text{(Eq. 5.7)}$$

It follows from Eq. 5.7 that if u_{CRM}^2 is equal to or less than $0.1\left(s_b^2 + \frac{s_w^2}{n}\right)$, then the uncertainty of the CRM makes a negligible contribution to the calculation of σ and may be ignored. This condition will be fulfilled when u_{CRM} is equal to or less than one third of the estimated standard uncertainty $\sqrt{s_b^2 + \frac{s_w^2}{n}}$, of the laboratory's measured result for the CRM (see Section 3.10).

Worked example

The CRM LGC7131 is a soft drink containing certified levels of artificial sweeteners. The certified concentration for saccharin is 76 mg L^{-1}, with an expanded uncertainty ($k = 2$) of 7 mg L^{-1}.

The results of six replicate measurements on the material are shown in Table 5.3. The value adopted for s_b is 1.5 mg L^{-1}, based on the long-term repeatability of the method.

The value for σ (based on Eq. 5.4) is given by:

$$\sigma = \sqrt{1.5^2 + \frac{0.44^2}{6}} = 1.51$$

The calculated value for the apparent bias ($\Delta = \bar{x} - \mu$) is $81.23 - 76 = +5.23$ mg L^{-1}.

It is seen that this value does *not* meet the requirements of Eq. 5.3, in that ($\bar{x} - \mu$) (5.23 mg L^{-1}) is *greater* than 2σ (3.02 mg L^{-1}). Based on these figures, therefore, we must conclude that, at the 95% confidence level, there *is* evidence for the presence of bias in the method.

However, comparison of the estimated value (1.51) for σ with the standard

Table 5.3

Replicate 1	81.46 mg L^{-1}
Replicate 2	81.64 mg L^{-1}
Replicate 3	80.47 mg L^{-1}
Replicate 4	81.59 mg L^{-1}
Replicate 5	81.25 mg L^{-1}
Replicate 6	80.99 mg L^{-1}
n	6
mean	81.23 mg L^{-1}
s_w	0.44 mg L^{-1}

uncertainty for the certified concentration value shows that the latter cannot be ignored.

$$u_{CRM} = 7/2 = 3.5 \text{ mg L}^{-1}$$

Rather than u_{CRM} being less than one third of the value of σ estimated as above, it is actually greater than σ. We therefore need to calculate σ using Eq. 5.6, rather than Eq. 5.4. The appropriate value for σ is therefore given by:

$$\sigma = \sqrt{1.5^2 + \frac{0.44^2}{6} + 3.5^2} = 3.81 \text{ mg L}^{-1}$$

It is now seen that the observed value for the apparent bias (5.23 mg L^{-1}) *does* comply with the requirements of Eq. 5.3, in that 5.23 is now less than the 2σ value ($2 \times 3.81 = 7.62$ mg L^{-1}).

Taking into account the uncertainty of the documented value of the CRM, we therefore conclude that there is no evidence of bias in the method, as carried out by the laboratory.

5.4.3 What Is the Uncertainty of the Bias Estimate?

It will be appreciated that any estimate of a bias ($\Delta = \bar{x} - \mu$) will have an uncertainty associated with it. Where the application of Eq. 5.3 leads to the detection of a bias (*i.e.* Δ lies outside the limits of $\pm 2\sigma$), the value 2σ may be adopted as an estimate of the 95% confidence interval for Δ. This is because the value $\bar{x}$ can be expected to be distributed about the true bias with a standard deviation σ. However, should the standard uncertainty (u_{CRM}) of the CRM be significant in comparison to σ (*i.e.* $u_{CRM} > 0.3\sigma$), then the 95% confidence interval for Δ is given by $\sqrt{(2\sigma^2 + U_{CRM}^2)}$, where U_{CRM} is the expanded uncertainty (coverage factor, $k = 2$) of the reference value.

Worked example

Using the data from the previous example in Section 5.3.7, the 95% confidence interval for the measured bias of 5.23 mg L^{-1} is calculated to be 3.02 mg L^{-1}.

When the uncertainty of the certified value is taken into account, the 95% confidence interval for the measured bias of 5.23 mg L^{-1} is calculated to be 7.62 mg L^{-1}.

In passing, it will be noted that the range of the first confidence interval (5.23 ± 3.02) does not encompass zero, whilst the range of the second confidence interval (5.23 ± 7.62) does. This observation is another way of arriving at the conclusions presented in the previous example, namely that a bias *is* detected when the uncertainty of the CRM is ignored, but is *not* detected when the reference value uncertainty is taken into account.[9]

5.4.4 What Is the Bias Detection Limit?

It was stressed in the discussion in Section 5.3.5 that even if the requirements of Eq. 5.5 are met, we can only conclude that there is *no evidence* for the presence of bias. Bias could indeed be present, but at a level which the test is incapable of detecting. It is therefore necessary to have some way of estimating what this undetectable level of bias could be (*i.e.* what is the bias detection limit?) for the circumstances of any particular test of bias. Armed with this information the analyst is able to make a judgement as to whether the method of analysis being used is fit for purpose. The analyst will be able to conclude, at a specified confidence level (usually 95%), that any bias in the results on routine samples does not exceed the bias detection limit.

In estimating the bias detection limit, the objective is to identify what *minimum* value of a (biased) population mean would lead to an apparent bias being observed, with 95% confidence,[9,10] if the experiment to assess bias (*e.g.* seven replicate measurements) were to be repeated an infinite number of times.

To tackle this problem we firstly recall that if we apply Eq. 5.5 and find that, for *any particular* assessment of bias, the apparent bias ($\Delta = \bar{x} - \mu$) lies outside the limits of $\pm 2\sigma$, we are able to conclude, with 95% confidence, that a bias has been detected. However, we also need to establish, with 95% confidence, that the observed value ($\bar{x}$) is indeed drawn from a population whose (biased) mean is such that an apparent bias (Δ) would be detected in *any* bias assessment experiment carried out.

This dual requirement is illustrated in Figure 5.1. The bias detection limit is the difference between μ (the documented reference value) and μ_B (the mean of the biased population). The result ($\bar{x}$) of any particular bias assessment is distributed about the mean (μ or μ_B) with a standard deviation σ (as defined in

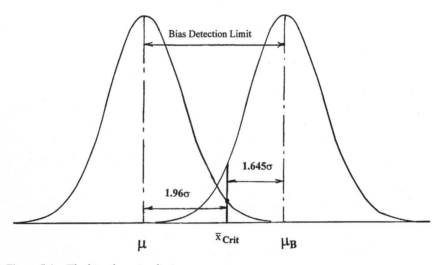

Figure 5.1   *The bias detection limit*

Eq. 5.4). The minimum bias detectable, with 95% confidence, is given by the value $\mu - \mu_B$ (see Figure 5.1).

The value $\bar{x}_{Crit}$ is located 1.96σ away from μ and represents the critical value of $\bar{x}$ that must be exceeded if we are to conclude, with 95% confidence, that a bias has been detected. The value 1.96 is taken from t-tables (95% confidence level, 2-tailed, ∞ degrees of freedom).

The value μ_B is located so that it differs from $\bar{x}_{Crit}$ by 1.645σ. The value 1.645 is taken from standard t-tables (95% confidence level, 1-tailed, ∞ degrees of freedom). It represents the requirement that the biased mean, μ_B, must be at least 1.645σ removed from $\bar{x}_{Crit}$, if we are to have 95% confidence that any measured value, $\bar{x}$, (from a population with a biased mean μ_B) will be greater than 1.96σ from μ in *any* experiment to assess bias. If this requirement is not included it would mean that whilst any one particular experiment to assess bias could well lead to a bias being detected ($\bar{x} > \bar{x}_{Crit}$), a subsequent experiment could (with a probability greater than 5%) lead to the conclusion that there was no evidence for bias ($\bar{x} < \bar{x}_{Crit}$).

The two-tailed t-value is applied to the distribution of a result $\bar{x}$ about μ, as we need to take account of the possibility of both positive and negative biases being present. However, once the nature of any bias has been established, *either* positive *or* negative, we need to consider only *one* half of the distribution curve of $\bar{x}$ about μ_B (the half that overlaps the distribution about μ). The one-tailed t-value is therefore applied to μ_B.[9,10] The t-values for infinite (∞) degrees of freedom are used as it is presumed that the value of σ is a reliable estimate of the measurement error, based on extensive experimental data. This assumption will usually be justified if any of the procedures 1 to 4 in Section 5.3.2 have been used to estimate σ, although in the case of procedure 1 seven replicate determinations are the recommended minimum and up to 20 are preferable (see Section 3.3).

The calculation of the bias detection limit (the minimum detectable bias), at the 95% confidence level, is therefore given by Eq. 5.8, where the t-values are rounded up to one decimal place:

$$\text{Bias detection limit} = (2.0 + 1.7)\sigma = 3.7\sigma \qquad \text{(Eq. 5.8)}$$

This equation enables us to calculate, with 95% confidence, the maximum bias that *could* be present in a particular method, when the measured result $\bar{x}$ obtained on a matrix reference material complies with the requirements of Eq. 5.5 and we conclude that there is *no evidence* for the presence of bias. The actual bias could be any value between 0 and the bias detection limit.

The foregoing discussion assumes that the uncertainty of the documented value of the CRM (the true value) is negligible compared with the estimated uncertainty of the measurement σ. If this is not the case, the uncertainty of the true value must be taken into account when calculating the bias detection limit. If the quoted expanded uncertainty for a coverage factor of 2 is U, the bias detection limit is calculated (with a 95% confidence level) using Eq. 5.9:

$$\text{Bias detection limit} = \sqrt{(3.7\sigma)^2 + U_{CRM}^2} \qquad \text{(Eq. 5.9)}$$

If the expanded uncertainty U is quoted for a coverage factor of 3, it may be readily converted to the corresponding value for a coverage factor of 2 by multiplying by 2/3.

It is seen from Eq. 5.9 that the uncertainty of the CRM places a lower constraint on the bias detection limit, in that the latter can never be smaller than the CRM's expanded uncertainty, U, (or smaller than twice the standard uncertainty, u).

Worked example
The CRM LGC7103 is a sweet digestive biscuit matrix with certified values for various nutritional components. The certified value for the chloride content is 0.54 g $(100\ \text{g})^{-1}$, with an expanded uncertainty ($k = 2$) of 0.07 g $(100\ \text{g})^{-1}$.

Duplicate analyses were carried out on the CRM, using a method with a known repeatability standard deviation (s_w) of 0.01 g $(100\ \text{g})^{-1}$. The results obtained were 0.49 and 0.51 g $(100\ \text{g})^{-1}$ (mean = 0.50 g $(100\ \text{g})^{-1}$). The long-term repeatability of the method (s_b) was estimated as 0.026 g $(100\ \text{g})^{-1}$, from previous data.

The value for σ (based on Eq. 5.4) is therefore calculated as:

$$\sigma = \sqrt{0.026^2 + \frac{0.01^2}{2}} = 0.027 \text{ g } (100\ \text{g})^{-1}$$

The apparent bias = 0.50 − 0.54 = −0.04 g $(100\ \text{g})^{-1}$.
This value complies with the requirements of Eq. 5.3, since:

$$-0.054 < -0.04 < +0.054.$$

It is therefore concluded that there is no evidence for the presence of bias.
To estimate the bias detection limit, we apply Eq. 5.8:

$$\text{Bias detection limit} = 3.7 \times 0.027 = 0.0999 = 0.1 \text{ g } (100\ \text{g})^{-1}$$

On this basis, it may be concluded, with 95% confidence, that the bias in the laboratory's application of this method does not exceed 0.1 g $(100\ \text{g})^{-1}$.

However, the expanded uncertainty U ($k = 2$) of the reference value is 0.07 g $(100\ \text{g})^{-1}$, which is not negligible compared to the term 3.7σ (0.1 g $(100\ \text{g})^{-1}$). Eq. 5.9 is therefore used to obtain a more realistic estimate of the bias detection limit:

$$\text{Bias detection limit} = \sqrt{0.1^2 + 0.07^2} = 0.12 \text{ g } (100\ \text{g})^{-1}$$

We conclude (with 95% confidence) that, based on this data, the laboratory's bias in the execution of the method does not exceed 0.12 g $(100 \text{ g})^{-1}$. The true bias could be any value between 0 and 0.12 g $(100 \text{ g})^{-1}$.

If it is considered that the *possible* presence of this level of bias $(0.12 \text{ g } (100 \text{ g})^{-1})$ is too high, the laboratory would need to identify an alternative analytical method with a smaller σ, which effectively means a method with a smaller long term repeatability, s_b.

5.4.5 Why Is the Bias Detection Limit (3.7σ) Greater than the Acceptability Criterion of 2σ?

It will be noted that the bias detection limit of 3.7σ is greater than the criterion of 2σ in Eq. 5.5, which stipulates that a result that is 2σ removed from the reference value constitutes evidence of bias. At first sight this appears contradictory, but further consideration shows that it is not.[9,10] Whilst a result that is 2σ or more removed from the reference value can be held, with 95% confidence, to represent a bias it does *not* automatically follow that we can be 95% confident that the bias actually observed in the experiment is the true bias. The observed bias is actually derived from a population (with a standard deviation σ) whose mean μ_B represents the true bias. If, for example, an observed bias of 2σ (just sufficient to be significant in any one experiment to assess bias) was (by chance) equal to μ_B, it would be quite likely that a subsequent experiment would yield a result that was less than 2σ from the reference value. It would then have to be concluded that no bias had been detected. The 3.7σ criterion defines the (minimum) level of bias that could be consistently detected, with 95% confidence, in any bias assessment. Biases lower than 3.7σ, but greater than 2σ can be detected, but with a probability lower than 95%. Figure 5.1 may help to clarify the situation.

5.4.6 How May a Fitness-for-Purpose Criterion be Used to Calculate Acceptability Limits?

It should be appreciated that Eq. 5.5 defines the acceptability of a result on a matrix CRM entirely in terms of the analytical parameter σ. This represents the expected uncertainty in the result due to the method of analysis used. It takes no account of the accuracy actually required by the ultimate end-user of the analytical data. In some circumstances the value σ may place an unnecessarily stringent acceptability limit on the result $\bar{x}$ obtained on a CRM, which is not justifiable, either technically or economically. When the uncertainty that is acceptable in a result is greater than the uncertainty that is analytically possible, a modified form of Eq. 5.5 may be used to define a wider acceptability limit for the value $\bar{x} - \mu$. The modified equation is:[3]

$$-a_2 - 2\sigma < \bar{x} - \mu < 2\sigma + a_1 \qquad \text{(Eq. 5.10)}$$

The analyst chooses the values a_1 and a_2, possibly after discussion with the end-user of the data, to reflect the uncertainty that is acceptable in the results on routine samples.

Worked example

In a survey of soil from an industrial site for contamination by polychlorobiphenyls (PCBs), it was decided to carry out an initial rapid screening of a large number of samples. Any 'hot-spots' identified would then be studied in more detail. In order to carry out the initial screen in a cost-effective manner, it was agreed that an uncertainty of $\pm 50\%$ in the results would be fit for purpose. This level of uncertainty would allow soil samples that differed in PCB content by a factor of 5 or more to be easily distinguished, *viz.*

nominal PCB content 10 mg kg^{-1}; observed result 10 ± 5 mg kg^{-1}

nominal PCB content 2 mg kg^{-1}; observed result 2 ± 1 mg kg^{-1}

For this application, the BCR certified reference material CRM481, an industrial soil with certified concentrations for various PCB congeners (*e.g.* PCB 118 certified concentration $= 9.4$ mg kg^{-1}) was used. The criterion for acceptable PCB 118 results on this material would therefore be 9.4 ± 4.7 mg kg^{-1}, or, using the format of Eq. 5.10:

$$-4.7 < \bar{x} - \mu < 4.7$$

Thus:

$$-a_2 - 2\sigma = -4.7 \quad \text{and} \quad a_1 + 2\sigma = 4.7$$

The contribution due to σ (*i.e.* the contribution due to the analytical method only) may be estimated in the usual manner, using Eq. 5.2. The within-laboratory repeatability standard deviation (s_w) for analyses at the nominal 10 mg kg^{-1} level is taken to be 0.5 mg kg^{-1}, whilst the estimate adopted for s_b, based on the Horwitz function, is 1 mg kg^{-1}. The value of σ, for a single analysis ($n = 1$) is therefore given by:

$$\sigma = \sqrt{1^2 + \frac{0.5^2}{1}} = 1.1 \text{ mg kg}^{-1}$$

Thus, ignoring the effect of the 'a' terms in Eq. 5.10, the acceptability limits would be:

$$-2.2 < \bar{x} - \mu < 2.2$$

Therefore, based on purely analytical considerations, a result on the CRM outside these limits would indicate that a bias had been detected and that corrective action should be initiated.

However, in the present example such action is seen to be unwarranted, since the analysis is only required to have acceptability limits of ± 4.7 mg kg^{-1}. Thus, although a bias has been detected if $-2.2 < \bar{x} - \mu < 2.2$, it does not constitute a genuine problem when the fitness for purpose limits of ± 4.7 are adopted. Corrective action would then only be necessary if a result on the matrix CRM differed from the certified value by more than the ± 4.7 limits.

The practical consequence of the above considerations is that the proposed analytical method ($s_w = 0.5$ mg kg^{-1} and $s_b = 1.0$ mg kg^{-1}) is more than fit for purpose. Therefore, it might be appropriate to adopt a somewhat less accurate procedure for the screening phase of the study that could be carried out more rapidly and/or at lower cost.

5.5 Action to Take When a Bias Is Detected

If, following the analysis of a matrix CRM, the application of Eq. 5.3 shows that the result lies outside the acceptability limits it must be concluded that a bias has been detected in the laboratory's execution of the method. Appropriate remedial action is then required and some steps that may be taken are discussed below.

5.5.1 Check the Calibration Standards

It has been reported[7] that one of the most common causes of biased data is the use of poor quality and inaccurate instrumental calibration standards. A standard solution must be properly prepared, stored and maintained if it is to be used to provide reliable calibration data. If the calibration standard is wrong, all results based on that standard will also be wrong. Ideally, a certified reference material should be used to prepare the calibration standard, rather than a lower grade material whose property values (*e.g.* purity) will not have been so well characterised. Even when a certified reference material is used to prepare the calibration standard, it is important that the preparation is carried out carefully and that the prepared standard is stored properly, to ensure that there is no subsequent deterioration of the standard prior to use. Chapter 4, Sections 4 and 5 provide further discussion on the proper selection, preparation and use of instrumental calibration standards.

5.5.2 Check Other Possible Sources of Bias

In addition to calibration errors due to the use of an inaccurate calibration standard, the following checklist highlights some other common sources of bias in the execution of an analytical method:

- contamination
- chromatographic and other interferences
- instrumental drift
- uncorrected blank

- incomplete extraction, digestion *etc.*
- losses during sample preparation
- matrix effects
- operator bias
- incorrect reagents
- method not being followed correctly
- data transcription errors
- calculation errors
- method being used outside its recommended scope

A systematic study is generally required to establish the importance, if any, of these effects. The bracketing technique may be used to reduce the effects of instrumental drift (see Section 4.12), whilst procedures for dealing with matrix effects are discussed in Section 4.13.

5.5.3   Check the Precision of the Measurements

If the within-laboratory standard deviation (s_w) appears unduly large (*i.e.* s_w comparable to or larger than s_b), this suggests that there is some uncontrolled source of short-term random variation that requires investigation.

For this purpose it is not necessary to use a matrix CRM. A bulk sample that is similar to the matrix CRM and that is known to be homogeneous would be adequate for an experimental investigation aimed at improving the repeatability of the execution of an analytical method.

The following checklist provides some examples of common sources of poor precision in the execution of an analytical method:

- instrumental instability
- reagent variability
- variability of blank
- variable contamination from reagents, apparatus, *etc.*
- operator inexperience
- environmental fluctuations
- variable losses during sample preparation, extraction, digestion, *etc.*

A systematic study is generally required to establish the importance, if any, of these effects. If instrumental instability is suspected, the bracketing technique for calibration (see Section 4.12) may be beneficial.

To test whether an observed value of s_w is acceptable it may be compared to the expected value using the Chi-squared test (see Section 3.5). The expected value could be based on the repeatability standard deviation (s_r) quoted in the documentation for the method, where the method has been the subject of an inter-laboratory validation exercise conducted according to recognised protocols. Alternatively, the expected value could be based on the laboratory's previous experience of the procedure. This approach may be particularly useful where a less experienced operator has obtained a biased result.

5.5.4 Apply a Correction Factor?

One possible approach to dealing with results that are known to be subject to bias would be to apply a correction factor to the results to compensate for the bias. However, this approach is not generally recommended, except as a 'last-resort' and even then the use of a correction factor should be made clear when the results are reported. The uncertainty of the correction factor should be incorporated into the estimate of the total uncertainty of the results. It is far preferable to identify the source of bias in the execution of the method and to take appropriate remedial action at that point.

5.6 References

1. *The Fitness for Purpose of Analytical Methods. A Laboratory Guide to Method Validation and Related Topics*, EURACHEM, 1998.
2. Protocol for the Design, Conduct and Interpretation of Method Performance Studies, *Pure Appl. Chem.*, 1995, **67**, No. 2, 331–343.
3. *Uses of Certified Reference Materials*, ISO Guide 33, 1989.
4. *Certification of Reference Materials – General and Statistical Principles*, ISO Guide 35, 1989.
5. *Quality Assurance of Chemical Measurements*, J. K. Taylor, Lewis Publishers Inc., 1987 (ISBN 0-87371-097-5). Chapter 24, Correction of Errors and Improving Accuracy.
6. *International Vocabulary of Basic and General Terms in Methodology*, 2nd edition, 1993 (ISBN 92-67-01705-1).
7. Quality Assurance in the Analysis of Foods for Trace Constituents, W. Horwitz, L. R. Kamps and K. W. Boyer, *J. Assoc. Off. Anal. Chem.*, 1980, **63**, No. 6, 1344–1354.
8. International Harmonized Protocol for Proficiency Testing of (Chemical) Analytical Laboratories, M. Thompson and R. Wood, *J. AOAC Internat.*, 1993, **76**, No. 4, 926–940.
9. *Use of NIST Standard Reference Materials for Decisions on Performance of Analytical Chemical Methods and Laboratories*, NIST Special Publication 829, January 1992.
10. *Statistics for Analytical Chemistry*, 3rd edition, J. C. Miller and J. N. Miller, Ellis Horwood, 1993 (ISBN 0-13-030990-7). Chapter 3, Significance Tests, Section 13.

CHAPTER 6

Use of CRMs in Method Validation and Assessing Measurement Uncertainty

6.1 What Is Method Validation?

The ISO definition of validation[1] is 'confirmation by examination and provision of objective evidence that the particular requirements for a specified end-use are fulfilled'. There are alternative definitions of validation, from other sources, but these are broadly all interpretations of the ISO definition.

Validation has three important component parts and, when applied to method validation, these translate into:

- the specified end-use is the analytical requirement which derives from the problem that the analysis is intended to solve
- the objective evidence is usually in the form of data from planned experiments, from which the appropriate method performance parameters are calculated
- the confirmation is taken as a satisfactory comparison of the performance data with what is required, *i.e.* the method is fit-for-purpose

Thus, method validation can be seen to be a process of developing a sufficient picture of a method's performance to demonstrate that it is fit for an intended purpose. The process is based on determination of a range of performance characteristics of the method. Some are quantitative (trueness [bias], precision, detection limits); others qualitative or semi-quantitative (ruggedness, selectivity). In Figure 6.1 the components of a method validation study are represented as a jigsaw puzzle.

Though it is safe to say that a validation is most unlikely to be adequate without some study of overall precision and bias, there is no entirely general order of preference; different circumstances will place emphasis on different aspects of the process. Most jigsaws yield a recognisable picture without all the pieces; validation is no exception. The analogy can be extended somewhat: substantial effort will provide a more sharply focussed picture than a short

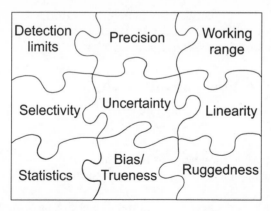

Figure 6.1 *Method validation puzzle*

study. It inevitably falls to the analyst to decide, with sectoral guidance if necessary, what parts of the puzzle are essential and how sharp a focus is necessary. A validation plan specifies both which parts of the puzzle must be assembled and the degree of focus, or effort, required.

Advice on how to do method validation is laid out in a number of guides – the actual procedures may vary from sector to sector and a comprehensive Eurachem Guide on method validation has recently been published.[2] It is always worth following any guidance available for your particular sector of work, so that your validation is compatible with that in peer laboratories. Where particular conventions have been followed for particular method validation calculations these should be stated.

6.2 Why Is Method Validation Necessary?

Method validation is necessary for a number of reasons. It is an important element of quality control. Before a measurement is carried out there should be some assurance that it will be correct. Validation helps to provide that assurance. It also provides data which are used as the basis for comparisons in quality control. The unsatisfactory alternative is to carry the measurement out, detect errors and have to repeat it. (It is cheaper/better to prevent problems from happening rather than have to correct them afterwards). In a production environment the producer has a responsibility to have taken all reasonable care in the quality of a product before releasing it to the consumer. Validation provides part of the minimum liability. In some areas, the validation of methods is a regulatory requirement.

An offshoot of validation is that in measuring the method performance parameters, data will be accumulated showing which parts of the method are stable and which can cause problems in overall performance. This, in turn, enables suitable quality control procedures to be designed and implemented. Method validation data provide information which enables the comparability of results from samples analysed in different laboratories and using different methods to be assessed.

Overall, the validation of methods is good science.

6.3 When Do You Validate a Method?

Method validation is most commonly carried out in the following circumstances:

- during method development (own method)
- before using any method for sample analysis (own, published or standard method)
- revalidation for changes to application or working environment or following long periods of non-use (own, published or standard method)

Validation usually begins during the method development stage. During the development of a method, performance parameters will be evaluated approximately. This will give an idea as to whether the method capabilities are in line with the levels of capability required. Once the method is deemed good enough, the development phase finishes, giving way to more formal validation studies.

Published methods may not necessarily be properly validated. The analyst is always advised to check the level of validation against that required and add further validation as required. The analyst who uses the method routinely will not necessarily be the same one who has carried out the validation. Methods are sometimes validated in one part of a laboratory and then transferred to other parts for routine use.

Whether the validated method is published or has been developed in-house, the analyst who uses it should first confirm that the validation data and the method's fitness-for-purpose applies to the samples being analysed. This is sometimes known as verification of performance. A change of use of the method requires the validation to be checked. Extending the use of the method to different sample types, or analyte levels, will require the performance to be rechecked using the new type of samples. The effect of changes to other parameters such as analyst, instrument, laboratory environment should also be checked.

6.4 How Do You Validate a Method?

Method validation is not an accidental activity, it should be both deliberate and planned.

The following steps will normally be undertaken in validating a method:

- decide analytical requirements
- plan a suite of experiments
- carry out experiments

- use data to assess fitness for purpose
- produce a statement of validation

The first stage is to examine the problem presented by the customer. Look at the reasons behind carrying out the analysis and find out what it is that the customer hopes to establish from having the work carried out. From this it should be possible to decide which method performance parameters are relevant to the work and what sort of target values are required. From this a suite of experiments can be designed which should gather relevant data. The plan will include details on what is going to be analysed at each stage, what degree of replication is required. It is possible that several parameters may be examined in one set of experiments, in which case the order in which things are done can be important. Once the plan is finalised the parameters are evaluated and the data used to decide whether the method is fit for purpose. The statement of validation is the positive assertion of fitness for purpose. Depending on how much work is done the statement may simply apply to the particular problem – 'yes the method is fit for purpose'. Alternatively, a more open ended validation may be added to the scope of the method, stating fitness for purpose for types of matrices, analytes, levels, degree of precision and accuracy, *etc.*

6.5 Method Performance Parameters

The following method performance parameters are the most important for the vast majority of analytical methods:

- confirmation of identity (selectivity/specificity)
- trueness (bias, recovery)
- precision (repeatability, reproducibility)
- working range
- ruggedness/robustness
- sensitivity

Different method performance parameters will be important in different situations. Trueness may be important for calculating absolute values of properties or analytes. Precision is important in comparative level studies. Working range will be of interest in most cases. For trace work, limits of detection and quantitation may be relevant. For planning calibration strategies it may be useful to know over what range the response is linear. Ruggedness studies, carried out mainly during method development, will indicate which parameters need to be controlled in order to preserve performance. This in turn enables suitable quality control strategies to be devised. The meaning of *sensitivity* depends on the sector in which it is used. In an instrumental sense it refers to the response per rate of change of analyte/property value. Medical and clinical chemists use it as an alternative to *limit of detection*.

6.6   The Tools of Method Validation

A number of tools are available to the analyst designing experiments to determine method performance parameters including:

- standards and CRMs
- blanks
- reference materials prepared in-house and spikes
- existing and incurred samples
- statistics
- common sense

Most of the experiments are likely to involve some form of replicate analysis of samples that either contain none of the analyte of interest (blanks) or a fixed known amount. The latter may simply be standard solutions of the analyte, to monitor response related factors, or matrix-based samples which test the method's performance in the presence of the sample matrix or other known interferences. The replicated data are then analysed using statistics.

To monitor bias and recovery, ideally samples in which analyte levels are well characterised, *e.g.* matrix CRMs, are necessary. Pure analyte standards do not really test the method in the same way that matrix based samples do. Matrix CRMs are ideal if available. If not available, then a reference material prepared in-house is the next best option. The role of reference materials in method validation is covered in detail later in this chapter.

In the absence of suitable reference materials, it is possible to carry out recovery on spiked samples. However, the analyte tends to be bound less closely in spiked samples than in real samples and consequent recoveries tend to be over-optimistic. Incurred samples are those in which the analyte has been introduced at an early stage and is consequently realistically held in the matrix. Analyte levels may also have been affected by metabolism. The value of incurred samples very much depends on how well the analyte level can be characterised.

Each validation situation is unique and different strategies may be involved. Overall the best tools that the analyst can apply are their professional knowledge and common sense.

As mentioned previously, method validation begins during the method development stage. This should yield some approximate pointers that the method is likely to be fit-for-purpose. A more thorough validation regime (probably characterised by a higher degree of replication, and investigation of more concentration levels) is applied to the method when the initial development stage is over. A number of assumptions related to external factors are made at this point:

- The equipment on which the work is being done is broadly suited to the application. It is clean, well-maintained and within calibration.

- The staff carrying out the validation are competent in the type of work involved.
- There are no unusual fluctuations in laboratory conditions and there is no work being carried out in the immediate vicinity that is likely to cause interferences.
- The samples being used in the validation study are known to be sufficiently stable.

Provided that these assumptions hold, one can expect to obtain method performance data that are typical for the method when it is used under normal laboratory operating conditions.

Once the method performance data is gathered, an assessment can be made of whether the required target values for the method are met. If they are achieved, then the method can be declared as fit-for-purpose and considered to be validated. If the target values are not achieved further development of the method will be necessary, followed by reassessment of the target values.

6.7   The Role of CRMs in Method Validation

The main use of certified reference materials in method validation is to assess the trueness (bias) of a method, although with careful planning of experiments other useful information including method precision can be collected at the same time.

6.7.1   What Is Meant by Method Accuracy, Bias and Trueness?

In order to better understand the *bias* associated with an analytical method, it is useful to give an explanation of the terms random and systematic error and bias (trueness).

It is easy to see random effects; every result is different. Random effects are assessed using *precision* experiments. But experimental work is invariably subject to systematic effects, too. A method can accordingly be considered 'validated' only if the systematic effects are duly considered. 'Due consideration' may show that bias study is unimportant; direct comparisons within a short timescale will be unaffected by most systematic effects. But for most purposes, some bias check is relevant.

Systematic effects are associated with the property of *trueness* of a measurement, and may lead to *biased* results. However, it should be noted that under the current ISO definitions, *accuracy* is a property of a result, and comprises both bias and precision.

The ISO definition of *trueness*[3] states 'the closeness of agreement between the average value obtained from a large set of test results and an accepted reference value' with a note that 'the measure of trueness is normally expressed in terms of bias'.

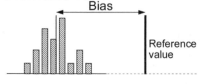

Bias

- Difference between observed mean value and reference value

- Bias is a measure of *Trueness*

Figure 6.2 *A visual illustration of analytical bias*

6.7.2 Measuring Bias Against CRMs

Tests to measure bias need:

- sufficient precision to detect practically significant bias (*i.e.* the maximum acceptable bias for the method to be fit for purpose)
- use of the most appropriate reference materials and certified values available
- tests covering the scope of the method adequately (*i.e.* range of analyte concentrations and matrices specified in the scope)

Two things immediately follow from the definitions of bias given:

1. Any measure of bias should constitute an *average* reading.
2. A test for bias must be made on a test item with a known or accepted reference value, *e.g.* a CRM.

In general, an average value is compared with a reference value using a statistical *t*-test (see Section 6.7.5). Such tests become more sensitive as the number of measurements increases. It is important to choose the right number of replicates. Advice on the number of replicates to analyse is covered in Section 3.3 (p. 19), but in general a minimum of seven replicates, and up to 20 replicates is recommended if time and resources permit. The reference value used needs to be appropriate for the job. Ideally, it will be a certified reference value; if not, best attempts within available resources may include comparison with reference methods, study of in-house reference materials, spiking or other studies.

Finally, bias checks should cover the full scope of the method – analyte levels and matrices in particular – to check that bias is sufficiently controlled across the full range of measurements intended.

In planning a bias experiment, it is important to know how large a bias needs to be in order to be detected, and how precise the method is in practice. Sufficient precision for bias checks depends on having enough replicates and appropriate controlled analytical conditions. In method validation, it is normally sufficient to check that the bias is small compared with the standard deviation of single results. But, if results are subsequently averaged on the assumption of negligible bias, the result as used is an average, and averages are more precise than single values. In such circumstances, it will be necessary to control bias more closely; typically, the bias needs to be small compared with the precision of the results as used. Bias checks will often be conducted under repeatability conditions. This usually results in the bias check being more sensitive, but it is possible to find the method apparently unbiased for the short period in question. To be safe, the in-house reproducibility should also be checked.

6.7.3 How Bias is Normally Expressed

Bias can be expressed in one of two ways:

- As an absolute value, *i.e.* $x - x_0$, where a positive bias means a higher observed value

or (more commonly in method validation) as a recovery factor

- *i.e.* as a fraction or percentage, x/x_0 or $100x/x_0$

where x is the observed value and x_0 is the reference value.

Bias for a single material is usually expressed as a simple difference between observed and expected result. The difference is usually taken so that a higher observed result gives a positive bias. In some cases, it is more useful to describe bias in terms of a ratio. For example, analytical recovery is usually expressed as a percentage. This form is most useful when many tests or materials are subject to the same proportional bias.

6.7.4 Measuring Bias Against a Reference Method

Often, methods are developed to substitute for a more costly 'standard method'. For example, combustion methods are quicker and easier than the digestion/titration Kjeldahl method for nitrogen determination. Where possible, any method used for regulation should be checked for bias against a suitable CRM. However, in the case of a standard method with which the laboratory is competent, it is useful and sensible to compare the two methods directly.

The principle of the comparison of methods is that a number of samples are run using each method, forming a series of pairs of results. The differences are analysed statistically. If the methods perform identically, the mean difference (bias) would be zero, so the mean difference is tested for significant departure from zero (see Section 3.8).

Although direct comparison of methods does not require any particular material, it is preferable to use CRMs because apart from having certified values with a stated level of uncertainty they also guarantee a level of homogeneity. Within-sample inhomogeneity needs to be minimised; inhomogeneous materials will give poor apparent precision and weaken the test. Note that a close match between two methods does not necessarily assure accuracy; at least one needs to be checked independently with CRMs for bias, where possible.

6.7.5   Statistical Treatment of Results from a Study of Bias: Comparison with a Certified or Reference Value

As noted out in Section 3.4, and for the statistics to be meaningful, all data should be studied first for outliers, suspect values or trends before being analysed statistically.

When conducting a bias experiment comparing the certified value for a reference material with the results obtained with the particular analytical method, we use the mean $\bar{x}$, and standard deviation, s, of n replicates, and use a t-test to compare the mean with the reference value. The bias experiment will have given a set of n observations $x_1, x_2 \ldots x_n$ on the reference material, which has a certified value x_0 and the general form of the equation used to calculate the t-value is as follows:

$$t = \frac{\bar{x} - x_0}{s/\sqrt{n}}$$ (Eq. 6.1)

If $t > t_{\text{crit}}$, the bias is statistically significant.

Worked example
This worked example illustrates the use of a certified reference material to evaluate analytical bias (recovery) for a method. The results of repeat experiments to measure the recovery of cholesterol from a certified reference material (BCR CRM 164) are shown in Table 6.1 below.

The certified value of BCR CRM 164 is 274.7 ± 9.0 mg $(100 \text{ g})^{-1}$ cholesterol (the uncertainty is given as an expanded uncertainty using $k = 2$, which corresponds approximately to a 95% confidence interval).

Having obtained actual experimental data for CRM 164 using the method under investigation, the next steps are to consider:

1. What statistical test(s) should be applied to the data?

Table 6.1 *Results of analysis of milk fat CRM 164*

Experiment number	Cholesterol [mg $(100\ g)^{-1}$]
1	271.41
2	266.31
3	267.78
4	269.55
5	268.74
6	272.53
7	269.53
8	270.12
9	269.65
10	268.59
11	268.42
Mean ($\bar{x}$)	269.33
Sample sd (s)	1.69
n	11
sd of the mean ($s/\sqrt{n}$)	0.51

2. Does the experiment show a statistically significant difference from the certified value of the reference material?
3. Is the bias practically significant?

1. What statistical test(s) should be applied to the data?

The experiment determines apparent analyte levels against a stated value. A *t*-test for significant difference between the observed mean and the reference value is therefore appropriate.

2. Does the experiment show a statistically significant difference from the certified value of the reference material?

Carrying out a manual *t*-test:

- Select a level of significance   95% ($\alpha = 0.05$)
- Decide number of tails   2 (the direction of the difference between the observed mean and the reference value is unimportant)
- Calculate degrees of freedom $v = n - 1 = 10$
- Look up the critical value $t_{crit} = 2.23$ (95% two tailed, $v = 10$)
- Calculate the test statistic (Eq. 6.1) $t = (274.7 - 269.33)/0.51 = 10.5$
- If $t > t_{crit}$, declare the result significant SIGNIFICANT

Since the calculated value of t is about five times t_{crit}, there is certainly a highly significant difference between the observed mean and the reference value. If using software, check the calculated *p*-value. Values below 0.05 indicate significance (at the 95% level). Chapter 3 gives more detailed information on *t*-tests.

3. Is the bias practically significant?

Consider the assigned uncertainty in the reference value; 274.7 ± 9.0 mg $(100 \text{ g})^{-1}$. The observed bias is well inside the reference material uncertainty. There is accordingly no reason to act on the apparent bias.

Note: With the reference material used in the bias (recovery) experiments, bias below 9 mg $(100 \text{ g})^{-1}$ would not have been convincingly detectable in any case. If that is too large to ignore, a different certified reference material with a smaller uncertainty about the reference value is required.

6.7.6   Statistical Treatment of Results from a Study of Bias: Comparison with a Reference Method

When conducting an experiment to compare the results obtained from the method being evaluated, with those obtained from a reference method (see Section 6.7.4), we use a *t*-test to compare the mean difference between the pairs of results with zero. The bias experiment will have given a set of *n* pairs of observations $(x_1, y_1), (x_2, y_2) \ldots (x_n, y_n)$. Calculate the mean, $\bar{d}$, and standard deviation, $s(d)$, of the *n* differences $d_i = x_i - y_i$. The general form of the equation used to calculate the *t* value is then as follows:

$$t = \frac{|\bar{d}|}{s(d)/\sqrt{n}} \qquad \text{(Eq. 6.2)}$$

Worked example
This example illustrates the use of in-house reference materials to compare two different methods. Table 6.2 gives a series of analyses performed on different samples using two different methods. The methods are for determining the phosphorus content of detergents; the reference method uses a filtration step. Each sample was divided into two parts: one half was filtered, the other half was not. The rest of the analysis was identical for both halves of the sample. The results from six different materials are displayed in Table 6.2.

Having obtained actual experimental data using the two methods under investigation, the next steps are to consider:

1. How should the data be evaluated?
2. Is there a statistically significant difference between the methods?
3. Is there a practically significant difference between the methods?

Table 6.2   *Results of analysis comparing two methods*

Sample	Filtered (A)	Unfiltered (B)	Difference (B − A)
A	37.1	34.7	−2.4
B	26.4	25.5	−0.9
C	26.2	25.2	−1.0
D	33.2	32.3	−0.9
E	24.3	24.2	−0.1
F	34.7	32.6	−2.1
Mean, d			−1.2
Standard deviation, s			0.86
n			6
$s/\sqrt{n}$			0.35

1. How should the data be evaluated?

The data are for six different samples, which are not expected to be similar. The differences between samples could overwhelm a real systematic difference between methods, so a t-test for the means of the two sets of data may be misleading.

However, the results are 'paired', so an alternative is available – the *paired t-test*. This is simply a t-test on the observed *differences*. If the methods perform identically, the difference between them should average zero. It is therefore sensible to perform a test on the observed differences, using the hypothesis 'the true difference is zero' against the alternative hypothesis 'the true difference is not zero'.

Note: Had the methods been compared using replicate analyses of the same sample, a t-test for difference between the two means would be appropriate.

2. Is there a statistically significant difference between the methods?

Following the usual sequence:

- Select a level of significance   95% ($\alpha = 0.05$)
- Decide number of tails   2
- Calculate degrees of freedom $v = n − 1 = 5$
- Look up the critical value $t_{crit} = 2.57$ (95% two tailed, $v = 5$)
- Calculate the test statistic (Eq. 6.2) $t = 1.2/0.35 = 3.52$
- If $t > t_{crit}$, declare the result significant   **SIGNIFICANT**

The methods therefore differ significantly at the 95% level of significance. More detailed information on t-tests can be found in Section 3.8.2.

3. Is the bias practically significant?

Whether this bias is acceptable depends on the use to which the method is put. Given the range of samples studied, an apparent bias of -1.2 (about 4% of the mean analyte level), and precision (judged from the standard deviation of differences) of the order of 0.5–1 the bias would certainly cause some samples to be considered different if analysed by the two methods. It is unlikely that a manufacturer would permit a systematic error of 4% in an important component, for example. However, if the level is checked simply to ensure compliance with a much higher upper limit, a bias of about 4% might well be acceptable.

6.7.7   Using Bias Information

The following summary should be born in mind when assessing and using bias information:

- Bias measures trueness.
- Experiments are based on comparison with certified reference values or methods.
- Study a representative range of analyte levels and matrices, *etc.*
- Interpret results objectively using *t*-tests.
- Insignificant bias found does not mean 'none'.

Bias information gathered during method development and validation is primarily intended to inform further method development and study. If a significant effect is found, action is normally required to reduce it to insignificance. Typically, further study and corrective action would be needed to identify the source of the error and bring it under control. Corrective action might involve, for example, changes to the method, alterations to a written protocol or additional training.

In some cases, however, bias can realistically be corrected for. It is, at present, unusual to correct an entire analytical method for observed bias; corrections may, however, be applied for known effects such as temperature. However, some analytical protocols require correction for observed extraction recovery error to ensure direct comparability of results between methods. There is no current consensus on correction for recovery; current IUPAC guidelines on the topic[4] recognise a rationale for either standardising on an uncorrected method or adopting a corrected method, depending on the end use.

6.8   Use of CRMs in Estimating Method Recovery and Measurement Uncertainty

This section illustrates the use of CRMs and analysis data from other samples in the validation and uncertainty evaluation of a method to determine vitamin A and vitamin E in infant formula. A worked example illustrates how to calculate various components of the uncertainty budget for method bias and method precision, and then goes on to show how to combine them to give a combined standard uncertainty and an expanded uncertainty for the method as a whole. The use of method validation data in uncertainty estimation is discussed in detail elsewhere.[5,6]

6.8.1   Outline of Method for the Determination of Vitamin A and Vitamin E in Infant Formula

A method was developed for the determination of tocopherol, retinol and carotene isomers in a range of food stuffs. This example considers the validation and calculation of an uncertainty estimate specifically for the determination of all-*trans*-retinol (vitamin A) and α-tocopherol (vitamin E) in infant formula. The homogenised sample is hydrolysed to release the retinol and tocopherol isomers. These are then extracted into mixed ethers and the extract concentrated. Portions of the extract are chromatographed on different HPLC systems to separate and quantify the isomers required. A normal phase system is used for the determination of α-tocopherol whilst a reversed phase system is used for the determination of all-*trans*-retinol. In each case calibration is by means of a single standard prepared by serial dilution of a stock solution.

6.8.2   Precision Study

A survey of the nutritional labelling of commercially available infant formulas indicated that that they were all of broadly comparable composition (fat, protein and carbohydrate levels) and contained similar concentrations of vitamin A and vitamin E. Three samples were chosen to cover the widest possible range of vitamin concentrations. Each sample was analysed a total of four times, in four separate extraction and HPLC runs. For each HPLC run, fresh calibration standards and mobile phase were prepared. In addition, a certified reference material was analysed in replicate for the estimation of $\bar{R}_m$ (estimate of mean method recovery obtained from the analysis of a CRM (see Section 6.8.3). These results were included in the precision study. The results are summarised in Table 6.3.

Table 6.3   *Summary of results from precision studies on the determination of all-*trans-*retinol and* α-*tocopherol*

	All-*trans-retinol*				α-*tocopherol*			
Sample	*Mean* *(mg kg^{-1})*	*sd* *(mg kg^{-1})*	*RSD*	*n*	*Mean* *(mg kg^{-1})*	*sd* *(mg kg^{-1})*	*RSD*	*n*
Infant formula A	9.67	0.146	0.0151	4	64.08	2.411	0.0376	4
Infant formula B	10.40	0.247	0.0238	4	69.05	2.156	0.0312	4
Infant formula C	6.67	0.217	0.0325	4	171.93	7.959	0.0463	4
SRM 1846	5.32	0.285	0.0536	6	278.51	10.663	0.0383	6

The standard deviations observed for all-*trans*-retinol were of a similar order of magnitude. This indicated that, within the range of analyte concentrations studied, the standard deviation is independent of analyte concentration. In this case, the uncertainty associated with method precision, $u(P)$, for samples with all-*trans*-retinol concentrations ranging from approximately 5 mg kg^{-1} to 10 mg kg^{-1}, was estimated as the pooled standard deviation. This was obtained using the following equation:

$$s_{pool} = \sqrt{\left(\frac{(3 \times 0.146^2) + (3 \times 0.247^2) + (3 \times 0.217^2) + (5 \times 0.285^2)}{3 + 3 + 3 + 5} \right)} = 0.238$$

(Eq. 6.3)

$u(P)$ for all-*trans*-retinol is therefore 0.238 mg kg^{-1}.

For α-tocopherol, the *relative* standard deviations (RSD) observed for the different samples were all of a similar order of magnitude. This indicated that the standard deviation is approximately proportional to analyte concentration, across the range studied. In such cases it is appropriate to pool the relative standard deviations to obtain the estimate of $u(P)$, using the following equation:

$$RSD_{pool} = \sqrt{\left(\frac{(3 \times 0.0376^2) + (3 \times 0.0312^2) + (3 \times 0.0463^2) + (5 \times 0.0383^2)}{3 + 3 + 3 + 5} \right)}$$
$$= 0.0387$$

(Eq. 6.4)

$u(P)$ for α-tocopherol is therefore 0.0387 as a relative standard deviation.

6.8.3   Bias (Recovery) Study and its Associated Uncertainty

Recovery is defined as the ratio of the observed result for the method to a reference value (see Section 6.7.5). The reference value can be, for example, the certified concentration of a reference material; the result obtained from an alternative definitive method; the concentration of a spike added to a sample matrix. In a 'perfect' method the recovery would equal unity. However, in reality, factors such as imperfect extraction or instrumental effects often result in a recovery which does not equal one. The experiments required to evaluate recovery and its associated uncertainty will depend on the scope of the method and the availability, or otherwise, of suitable CRMs. The contribution of recovery to the overall uncertainty budget for the method will depend on the significance of any difference in recovery from unity, and how it is treated (for example, whether or not a correction is applied). Any estimate of the uncertainty associated with recovery must cover the method scope. Therefore, a representative range of samples covering typical matrices and analyte concentrations must be considered. The recovery for a particular sample, R, can be considered as comprising three components:

- $\bar{R}_m$ is an estimate of the mean method recovery obtained from, for example, the analysis of a CRM or a spiked sample. The uncertainty in $\bar{R}_m$ is composed of the uncertainty in the nominal value (*e.g.*, the uncertainty in the certified value of a reference material) and the uncertainty in the observed value (*e.g.*, the standard deviation of the mean of replicate analyses).
- R_s is a correction factor to take account of differences in the recovery for a particular sample compared to the recovery observed for the material used to estimate $\bar{R}_m$.
- R_{rep} is a correction factor to take account of the fact that a spiked sample may behave differently to a real sample with incurred analyte.

These three elements are combined multiplicatively to give an estimate of the recovery for a particular sample, *i.e.* $R = \bar{R}_m \times R_s \times R_{rep}$. It therefore follows that the uncertainty in R, $u(R)$, will have contributions from $u(\bar{R}_m)$, $u(R_s)$ and $u(R_{rep})$. How each of these components and their uncertainties are evaluated will depend on the method scope and the availability of reference materials. In the simplest case the method scope covers a single matrix type and analyte concentration for which a representative CRM is available. However, the situation is often more complex than this. The method scope may cover a range of matrices and/or analyte concentrations and there may be no suitable CRM available. The evaluation of uncertainties associated with recovery is discussed in detail elsewhere.[7]

The approach to estimating recovery and the associated uncertainty for our worked example of the determination of all-*trans*-retinol (vitamin A) and α-tocopherol (vitamin E) in infant formula is discussed in detail in the following sections.

6.8.4 Estimating $\bar{R}_m$ and $u(\bar{R}_m)$ Using a Representative CRM

In this method validation study, a certified reference material, SRM 1846 [produced by the National Institute of Standards and Technology (NIST)], was available with a matrix and analyte concentrations representative of those which will be routinely analysed using the method. The CRM is an infant formula certified for a number of vitamins, including all-*trans*-retinol and α-tocopherol. Six portions of the CRM were analysed, each being taken through the whole analytical procedure. The results, together with the certified values for the material, are summarised in Table 6.4.

The Eurachem Guide on *Quantifying Uncertainty in Analytical Measurements*[8] may be consulted for more information on calculating standard uncertainties from reference material certificates.

The mean recovery $\bar{R}_m$ was calculated as follows:

$$\bar{R}_m = \frac{\bar{C}_{obs}}{C_{CRM}}$$ (Eq. 6.5)

Table 6.4   *Results from the analysis of infant formula SRM 1846*

	Certified values			Observed values	
Analyte	Concentration C_{CRM} (mg kg^{-1})	Quoted uncertainty (mg kg^{-1})[1]	Standard uncertainty $u(C_{CRM})$ (mg kg^{-1})[2]	Mean $\bar{C}_{obs}$ (mg kg^{-1})	Standard deviation s_{obs} (mg kg^{-1})
All-*trans*-retinol	5.84	0.68	0.35	5.32	0.285
α-Tocopherol	271	25	12.8	278.51	10.663

[1]The quoted uncertainty is an expanded uncertainty given at the 95% confidence level.
[2]The standard uncertainty is obtained by dividing the expanded uncertainty by 1.96.

where $\bar{C}_{obs}$ is the mean of the results from the replicate analysis of the CRM and C_{CRM} is the certified value for the CRM.

The uncertainty in the recovery, $u(\bar{R}_m)$, was calculated as follows:

$$u(\bar{R}_m) = \bar{R}_m \times \sqrt{\left(\frac{s_{obs}^2}{n \times \bar{C}_{obs}^2}\right) + \left(\frac{u(C_{CRM})}{C_{CRM}}\right)^2} \qquad \text{(Eq. 6.6)}$$

where s_{obs} is the standard deviation of the results from the replicate analyses of the CRM, n is the number of replicates and $u(C_{CRM})$ is the standard uncertainty in the certified value for the CRM.

Thus, for all-*trans*-retinol, using Eqs. 6.5 and 6.6:

$$\bar{R}_m = 5.32/5.84 = 0.911$$

$$u(\bar{R}_m) = 0.911 \times \sqrt{\left(\frac{0.285^2}{6 \times 5.32^2}\right) + \left(\frac{0.35}{5.84}\right)^2} = 0.0581$$

Similar calculations for α-tocopherol gave values of $\bar{R}_m = 1.028$ and $u(\bar{R}_m) = 0.0511$.

The above calculation provides an estimate of the mean method recovery and its uncertainty. The contribution of recovery and its uncertainty to the combined uncertainty for the method depends on whether the recovery is significantly different from 1 and, if so, whether or not a correction is made.

6.8.5   Estimating the Contribution of $\bar{R}_m$ to $u(R)$

Assuming an estimate of the recovery $\bar{R}_m$ and its uncertainty $u(\bar{R}_m)$ has been obtained, as described in Section 6.8.4, three possible cases arise:[4]

1. $\bar{R}_m$, taking into account $u(\bar{R}_m)$, is not significantly different from 1, so no correction to the analytical result is applied.

2. $\bar{R}_m$, taking into account $u(\bar{R}_m)$, is significantly different from 1, and a correction to the analytical result is applied.
3. $\bar{R}_m$, taking into account $u(\bar{R}_m)$, is significantly different from 1, but a correction to the analytical result is not applied.

To determine whether the recovery is significantly different from 1 a significance test is used. The test statistic t is calculated using the following equation:

$$t = \frac{|1 - \bar{R}_m|}{u(\bar{R}_m)}$$ (Eq. 6.7)

If the number of degrees of freedom associated with $u(\bar{R}_m)$ are known, compare t with the 2-tailed critical value, t_{crit}, for the appropriate number of degrees of freedom at 95% confidence. If t is less than the critical value then $\bar{R}_m$ is not significantly different from 1.

If the number of degrees of freedom associated with $u(\bar{R}_m)$ are unknown, for example if there is a contribution from the uncertainty in the certified value of a reference material, compare t with k, the coverage factor that will be used in the calculation of the expanded uncertainty (see Section 6.8.9 for guidance on selecting an appropriate value for k).

If $|1 - \bar{R}_m|/u(\bar{R}_m) < k$, the recovery is not significantly different from 1.
If $|1 - \bar{R}_m|/u(\bar{R}_m) > k$, the recovery is significantly different from 1.

Case 1

The significance test indicates that the recovery is not significantly different from 1 so there is no reason to correct analytical results for recovery. However, there is still an uncertainty associated with the estimate of $\bar{R}_m$ as the significance test could not distinguish between a range of values, *i.e.* there is a range of recoveries about 1 which would pass the test and indicate no significant difference. If the test statistic was compared with t_{crit}, the range is $1 \pm t_{crit}u(\bar{R}_m)$. The uncertainty associated with recovery in this case, $u(\bar{R}_m)'$, is given by:

$$u(\bar{R}_m)' = \frac{t_{crit} \times u(\bar{R}_m)}{1.96}$$ (Eq. 6.8)

If the test statistic was compared with the coverage factor k, the range is $1 \pm k \times u(\bar{R}_m)$. In this case the uncertainty associated with $\bar{R}_m$ is taken as $u(\bar{R}_m)$.

To convert to a relative standard deviation divide $u(\bar{R}_m)$ or $u(\bar{R}_m)'$ by the assumed value of $\bar{R}_m$. In this case $\bar{R}_m = 1$ so the standard deviation is equivalent to the relative standard deviation.

Case 2

As a correction factor is being applied, $\bar{R}_m$ is explicitly included in the calculation of the result. $u(\bar{R}_m)$ is therefore included in the overall uncertainty calculation as the term $u(\bar{R}_m)/\bar{R}_m$ as the correction is multiplicative.

Case 3

In this case, the recovery is statistically significantly different from 1, but in the normal application of the method, no correction is applied (*i.e.* $\bar{R}_m$ is assumed to equal 1). In principle, without the correction, we are ignoring a known bias. If the result is reported with its uncertainty, the range will not include the best estimate of the true value. A simple report of the result and its uncertainty is therefore likely to mislead the user.

The ISO approach to uncertainty estimation[9] does not allow for this situation as it is assumed that all known biases are corrected for. An acceptable approach, when the recovery is not corrected for, is to report the result and its uncertainty, together with R_m and its uncertainty $u(R_m)$. This allows the user to decide how to deal with the recovery issue, and to estimate the uncertainty should they decide to correct the result.

Where a separate report of the recovery is not feasible, IUPAC describes an alternative approach.[4] The uncertainty estimate is increased to take account of the fact that the recovery has not been corrected for. The increased uncertainty, $u(\bar{R}_m)''$, is given by:

$$u(\bar{R}_m)'' = \sqrt{\left(\frac{1 - \bar{R}_m}{k}\right)^2 + u(\bar{R}_m)^2}$$

(Eq. 6.9)

where k is the coverage factor which will be used in the calculation of the expanded uncertainty and $u(\bar{R}_m)''$ is expressed as a relative standard deviation by dividing by the assumed value of $\bar{R}_m$ as in Case 1.

Going back to our worked example, we must first calculate t.

For all-*trans*-retinol, using Eq. 6.7:

$$t = \frac{1 - 0.911}{0.0581} = 1.53$$

In this case, t was compared with the coverage factor, $k = 2$. As t is less than two there is no evidence to suggest that $\bar{R}_m$ is significantly different from 1. $\bar{R}_m$ was therefore assumed to equal 1 with an uncertainty, $u(\bar{R}_m)$, of 0.0581 (Case 1). A similar calculation for α-tocopherol also indicated that $\bar{R}_m$ was not significantly different from 1. $u(\bar{R}_m)$ was therefore estimated as 0.0511.

6.8.6 Calculation of R_s and $u(R_s)$ from Spiking Studies

Where the method scope covers a range of sample matrices and/or analyte concentrations, an additional uncertainty term is required to take account of differences in the recovery of a particular sample type, compared to the material

Table 6.5   *Summary of results from the study of R_s for the determination of all-trans-retinol and α-tocopherol*

	All-trans-retinol			α-Tocopherol		
Spiking level	Approx. target concentration (mg kg^{-1})	Mean recovery	n	Approx. target concentration (mg kg^{-1})	Mean recovery	n
1	14.0	0.981	4	84.0	1.028	4
2	18.0	0.991	4	99.0	1.103	4
3	22.0	0.996	4	114.0	1.167	4
	$u(R_s)$	0.00767		$u(R_s)$	0.0696	

used to estimate $\bar{R}_m$. This can be evaluated by analysing a representative range of spiked samples, covering typical matrices and analyte concentrations, in replicate. The number of matrices and levels examined, and the number of replicates for each sample, will depend on the method scope. Calculate the mean recovery for each sample. R_s is assumed to be equal to 1 (*i.e.* variations in sample matrix and analyte concentration have no effect on the recovery). The uncertainty associated with this assumption, $u(R_s)$, is given by the spread of mean recoveries observed for the different spiked samples. The uncertainty is therefore the standard deviation of the mean recoveries for each sample type.

In our worked example, R_s was estimated from spiking studies on infant formula A used in the precision study. It was not possible to produce a homogeneous bulk spiked sample so individual portions of the infant formula were spiked at the required concentration. Samples were prepared at the concentrations indicated in Table 6.5. Four samples were analysed at each concentration, in separate extraction and HPLC batches. The samples were spiked by adding solutions of all-*trans*-retinol and α-tocopherol to approximately 10 g of sample in a saponification flask, prior to the addition of any other reagents. The mean recovery for each sample was calculated using Eq. 6.10:

$$\bar{R}_m = \frac{1}{n}\sum_{i=1}^{n}\frac{C_{obs(i)} - C_{native}}{C_{spike(i)}} \qquad \text{(Eq. 6.10)}$$

where, $C_{obs(i)}$ is the concentration observed for sample i, $C_{spike(i)}$ is the concentration of the spike added to the sample i and C_{native} is the concentration of the analyte in the unspiked samples. C_{native} was taken as the mean of the results obtained from the precision study. The results are summarised in Table 6.5. $u(R_s)$ is the standard deviation of the mean recoveries obtained at each concentration.

6.8.7   Calculating R and $u(R)$

In the general case, the recovery for a particular sample, R, is given by $R = \bar{R}_m \times R_s \times R_{rep}$. However, since R_s and R_{rep} are generally assumed to equal 1, $R = \bar{R}_m$. The values of $\bar{R}_m$ and $u(\bar{R}_m)$ used depends on whether or not $\bar{R}_m$ is significantly different from 1, and if so, whether a correction to the result for a particular sample is applied (see also Section 6.8.5).

The uncertainty associated with R, $u(R)$ is given by:

$$u(R) = R \times \sqrt{\left(\frac{u(\bar{R}_m)}{\bar{R}_m}\right)^2 + \left(\frac{u(R_s)}{R_s}\right)^2 + \left(\frac{u(R_{rep})}{R_{rep}}\right)^2} \qquad \text{(Eq. 6.11)}$$

However, if $R_s = R_{rep} = 1$, the equation simplifies to:

$$u(R) = \bar{R}_m \times \sqrt{\left(\frac{u(\bar{R}_m)}{\bar{R}_m}\right)^2 + u(R_s)^2 + u(R_{rep})^2} \qquad \text{(Eq. 6.12)}$$

Since, for our worked example, a representative CRM was available for the estimation of $\bar{R}_m$ and $u(\bar{R}_m)$ for both all-*trans*-retinol and α-tocopherol, there is no need for the R_{rep} term. Both $\bar{R}_m$ and R_s are assumed to be equal to 1. R is therefore also equal to 1. $u(R)$ was calculated using Eq. 6.12 above.

Thus, for all-*trans*-retinol:

$$u(R) = \sqrt{0.0581^2 + 0.00767^2} = 0.0586$$

A similar calculation for α-tocopherol gives an estimate of $u(R) = 0.0863$. Note that, since in both cases $R = 1$, the uncertainty is the same whether expressed as a standard deviation or a relative standard deviation.

6.8.8 Calculation of Combined Standard and Expanded Uncertainties

Following the estimation of the individual components of the uncertainty using the procedures outlined earlier, the next stage is to combine the standard uncertainties to give a combined standard uncertainty for the method. How the individual uncertainty components are combined depends on whether or not they are proportional to the analyte concentration. If the uncertainty component is proportional to the analyte concentration then it can be treated as a relative standard deviation.

If, however, the uncertainty is fixed regardless of the analyte concentration then it should be treated as a standard deviation.

This leads to two possible cases:

1. All sources of uncertainty are proportional to the analyte concentration

In this case all the individual uncertainty components should be converted to relative standard deviations. For a result y which is affected by the parameters p,

$q, r \ldots$, which each have uncertainties $u(p)$, $u(q)$, $u(r) \ldots$, the uncertainty in y, $u(y)'$, is given by:

$$\frac{u(y)'}{y} = \sqrt{\left(\frac{u(p)}{p}\right)^2 + \left(\frac{u(q)}{q}\right)^2 + \left(\frac{u(r)}{r}\right)^2 + \ldots}$$ (Eq. 6.13)

2. Some sources of uncertainty are independent of analyte concentration

In such cases the uncertainty components that are independent of the analyte concentration must be combined as standard deviations. The uncertainty in the result due to parameters that are not concentration dependent, $u(y)''$, is given by:

$$u(y)'' = \sqrt{u(p)^2 + u(q)^2 + u(r)^2 + \ldots}$$ (Eq. 6.14)

To calculate the combined uncertainty in the result, $u(y)$, at a given concentration y, the concentration dependent and concentration independent uncertainties are combined as follows:

$$u(y) = \sqrt{(u(y)'')^2 + (y \times u(y)')^2}$$ (Eq. 6.15)

where $u(y)'$ is the combined concentration dependent uncertainties calculated using Eq. 6.13, expressed as a relative standard deviation and $u(y)''$ is the combined concentration independent uncertainty calculated using Eq. 6.14.

Note that when the uncertainty estimate is required for a single analyte concentration, the uncertainty components can be combined as either standard deviations or relative standard deviations; it will make no difference to the final answer.

Returning to our worked example, all-*trans*-retinol is an example of case 2 where some sources of uncertainty are independent of analyte concentration. Hence Eq. 6.15 is used to calculate the combined standard uncertainty for the method as a whole. Thus, $u(y)$, for the analyte all-*trans*-retinol is calculated using the equation:

$$u(y) = \sqrt{u(P)^2 + (y \times u(R))^2}$$ (Eq. 6.16)

where
$u(y)$ = combined standard uncertainty for all-*trans*-retinol at concentration y
$u(P)$ = uncertainty contribution from method precision for all-*trans*-retinol, *as a standard deviation*
y = concentration level of all-*trans*-retinol
$u(R)$ = uncertainty contribution from method recovery for all-*trans*-retinol, *as a relative standard deviation*

Thus, at a concentration of 5 mg kg^{-1},

$$u(y) = \sqrt{0.238^2 + (5 \times 0.0586)^2} = 0.377 \text{ mg kg}^{-1}$$

In our worked example, the analyte α-tocopherol is an example of case 1 where all sources of uncertainty are proportional to analyte concentration. Hence for α-tocopherol, the combined standard uncertainty for the method $u(y)$ is calculated by combining the individual uncertainties as relative standard deviations using Eq. 6.13 for two terms as below:

$$\frac{u(y)}{y} = \sqrt{\left(\frac{u(P)}{P}\right)^2 + \left(\frac{u(R)}{R}\right)^2} \qquad \text{(Eq. 6.17)}$$

where,

$u(y)$ = combined standard uncertainty for α-tocopherol at concentration y

$\dfrac{u(P)}{P}$ = uncertainty contribution from method precision for α-tocopherol, as a relative standard deviation

y = concentration level of α-tocopherol

$\dfrac{u(R)}{R}$ = uncertainty contribution from method recovery for α-tocopherol, as a relative standard deviation

Thus,

$$\frac{u(y)}{y} = \sqrt{0.0387^2 + 0.0863^2} = 0.095$$

Therefore, at a concentration of 280 mg kg^{-1}, $u(y) = (280 \times 0.095) = 26.6$ mg kg^{-1}

6.8.9 Expanded Uncertainty

The combined standard uncertainty obtained above must be multiplied by an appropriate coverage factor, k, to give the expanded uncertainty. The expanded uncertainty is an interval which is expected to include a large fraction of the distribution of values reasonably attributable to the measurand. For a combined standard uncertainty $u(y)$, the expanded uncertainty $U(y)$ is given by:

$$U(y) = k \times u(y) \qquad \text{(Eq. 6.18)}$$

The choice of coverage factor depends on knowledge of the use to which the result is put, the degree of confidence required and knowledge of the degrees of freedom associated with individual uncertainty components. For most purposes a coverage factor of $k = 2$ is recommended (however, see note below). For a normal distribution a coverage factor of 2 gives an interval containing approximately 95% of the distribution of values. For a higher level of confidence, k is chosen as 3. For a normal distribution a coverage factor of 3 gives an interval containing over 99% of the distribution of values.

Note: The use of coverage factors of 2 and 3 to give levels of confidence of approximately 95% and 99% respectively assumes that there are a reasonable number of degrees of freedom associated with the estimates of the major contributions to the uncertainty budget. For this, it is recommended that at least 10 determinations are carried out in the precision and trueness studies. If it is not possible to obtain this many replicates and either of these factors dominates the uncertainty budget, the coverage factor should be obtained from the table of critical values for the Student t test. For example, if the dominant contribution to the uncertainty budget was based on only four determinations this would give three degrees of freedom. The two-tailed t_{crit} value at the 95% confidence level is 3.182. It can therefore be seen that using uncertainty estimates based on only a small number of determinations will have a significant effect on the coverage factor and hence on the expanded uncertainty.

Returning to our worked example, with a combined standard uncertainty $u(y)$, the expanded uncertainty $U(y)$ is given by Eq. 6.18. Thus, for all-*trans*-retinol, the expanded uncertainty for the method $U(y)$, is calculated as follows:

$$\text{expanded uncertainty } U(y) = 2 \times 0.377 = \pm 0.754 \text{ mg kg}^{-1}$$

The coverage factor of 2 gives a level of confidence of approximately 95%.

For α-tocopherol the combined standard uncertainty was calculated as 26.6 mg kg^{-1} for a sample at a concentration of 280 mg kg^{-1}. The expanded uncertainty $U(y)$ is therefore:

$$U(y) = 2 \times 26.6 = \pm 53.2 \text{ mg kg}^{-1} \text{ at that concentration.}$$

The coverage factor of 2 gives a level of confidence of approximately 95%.

The above example illustrates the use of certified reference materials and analysis data from other samples in the validation and uncertainty evaluation of a method to determine vitamin A and vitamin E in infant formula. It shows how to calculate various aspects of the uncertainty budget (method bias and method precision) and then goes on to show how to combine them to give a combined standard uncertainty and an expanded uncertainty for the method as a whole. Precision and bias uncertainties form arguably the most important parts of the method validation puzzle referred to in Figure 6.1, but to complete the uncertainty budget one would also have to consider other sources of uncertainty which are beyond the scope of this guide. In the validation of the method to determine vitamin A and vitamin E in infant formula the other sources of uncertainty relate mainly to uncertainty in the extraction conditions and HPLC conditions.

It should be noted that in our worked example for all-*trans*-retinol the scope of validation is only for a concentration range from 5 to 10 mg kg^{-1} (the amount found in infant formula) and that extending the method for samples containing concentrations much outside this range would need further experiments to check method performance.

6.9 Summary

We stated earlier in the chapter that method validation is a process of developing a sufficiently detailed picture of a method's performance to demonstrate that it is fit for an intended purpose. The process is based on the determination of a range of performance characteristics of the method. Some are quantitative (bias, trueness, precision, detection limits); others are qualitative or semi-quantitative (ruggedness, selectivity), and the main role of reference materials in method validation is in estimating the bias (recovery) of a method and estimating its uncertainty.

Once the method performance data are gathered an assessment can be made of whether the required target values for the method are met. If they are achieved, then the method can be declared as fit-for-purpose and considered to be validated. If the target values are not achieved further development of the method will be necessary, followed by reassessment of the target values.

Once the validation plan is finalised the parameters are evaluated and the data used to decide whether the method is fit for purpose. The statement of validation is the positive assertion of fitness for purpose. Depending on how much work is done the statement may simply apply to the particular problem – 'yes the method is fit for purpose'. Alternatively, a more open ended validation may be added to the scope of the method, stating fitness for purpose for, for example, types of matrices, analytes, levels, degree of precision and accuracy.

6.10 References

1. *Quality – Vocabulary*, ISO 8402, 1994.
2. *The Fitness-for-Purpose of Analytical Methods, A Laboratory Guide to Method Validation and Related Topics*, EURACHEM, 1998 (ISBN 0-948926-12-0).
3. *Statistics – Vocabulary and Symbols – Part 1: Probability and General Statistical Terms*, ISO 3534-1, 1993.
4. M. Thompson, S. L. R. Ellison, A. Fajgelj, P. Willetts and R. Wood, Harmonised Guidelines for the Use of Recovery Information in Analytical Measurement, *Pure Appl. Chem.*, 1999, **71**, 337.
5. V. J. Barwick and S. L. R. Ellison, *Accred. Qual. Assur.*, 2000, **5**, 47.
6. V. J. Barwick and S. L. R. Ellison, *Accred. Qual. Assur.*, 2000, **5**, 104.
7. V. J. Barwick and S. L. R. Ellison, *Analyst (Cambridge)*, 1999, **124**, 981.
8. *Quantifying Uncertainty in Analytical Measurement*, EURACHEM, 1995 (ISBN 0-948926-08-2).
9. *Guide to the Expression of Uncertainty in Measurement*, ISO, 1993 (ISBN 92-67-10188-9).

Annex A   List of Reference Material Producers

Reference Material Producer	List of Products
1. Multi-sector	
IRMM, BCR Reference Materials Catalogue, 2000, European Commission, Belgium	wide ranging portfolio of materials
Laboratoire National d'Essais, France	wide ranging portfolio of materials
LGC Certified Reference Materials Catalogue, Issue No 6, 2000, UK.	wide ranging portfolio of materials
NIST Standard Reference Materials Catalog, 1998–1999, USA.	wide ranging portfolio of materials
NRCC, National Research Council, Canada	wide ranging portfolio of materials
2. Environment	
Accustandard	environmental pollution
Analytical Standards	steels, alloys, iron, ore, slags, rocks, ceramics, glasses, soils, muds, sediments, coals, cokes, pitches, oils
AQCS (Analytical Quality Control Services)	nuclear materials, sediment, soil, ores, biological
Brammer Standard Co. Inc	metals, glass, geochemical, slags, soils and sediments
British Ceramic Research Ltd	ceramics, geological samples
Cambridge Isotope Labs Ltd	stable isotopes, environmental contamination
CANMET	range of mineralogical and metallurgical matrices certified for elemental content, soils and sediments
Czechoslovak Atomic Energy Commission	environmental, dusts, ashes
Environmental Resources Associates	water, soil, organics/inorganics
Geological Survey of Japan (GSJ)	igneous rock (major and minor elements)
Glen Spectra	metals, coal, oils, minerals, ores *etc.*
International Atomic Energy Agency, Vienna	blood, bone, milk powder, flour, marine materials, fish, sea plants
National Institute for Environmental Studies (Japan)	pond sediment, exhaust particulate, rice flour, sargasso, chlorella, human hair, mussels

Reference Material Producer	*List of Products*
National Research Centre for CRMs, Beijing, China	metals, building materials, nuclear, polymers, chemical products, geology, environmental, clinical, food
National Water Research Institute, Canada	sediments, water
NOAA, National Oceanic and Atmospheric Administration, National Ocean Service, USA	reference materials for marine science
Powder Products Ltd	test dusts
Promochem Ltd	inorganic and organic standards
QMx	environmental solutions, pesticides, PAH, PCB, VOCs
Radian Corporation	environmental standards
Standard Laboratory of China, Environmental Monitoring Centre	water, soil
Trinidad & Tobago Bureau of Standards	fertilizers, pesticides, animal feed, household chemicals, building materials, textiles, metallurgy
Wzormat	pesticides, physicochemical, spectrometric

3. Industrial Products & Raw Materials

Alfa Products (Johnson Matthey, GmbH)	aqueous atomic absorption standards
Alusuisse-Lonza Services Ltd	steel and special alloys for spectrochemical analysis
Amt für Standarisierung, Messwesen und Warenprüfung (ASMW), Germany	metallurgy – certified steels, iron, gases in metals, spectrometric standards, metal content of oil, ion-activity, density, viscosity, radioactivity
Angstrom	metal organics, sulfur and chlorine organics, oils
ARRO Laboratory Inc	trace metal standards
BA Chemicals Ltd	alumina trihydrate flame retardants, thermosets, thermoplastics, elastomers, cellulosics
BAM (Bundesanstalt für Materialforschung und-prüfung)	steels, alloys, slags, ceramic materials
BDH	reagents, biochemicals, diagnostics, fine chemicals
BNF (British Nuclear Fuels)	metals, alloys, for spectrometric analysis
BOC	industrial gases
British Glass	'EC 1.1' float glass
Bureau of Analysed Samples Ltd (BAS)	metal alloys, ores, slags, ceramics, minerals, cement
Campro	labelled isotopes
Cartech	aluminium and aluminium alloys
Central Bureau for Nuclear Measurements, European Commission, Joint Research Centre	nuclear and isotopic reference materials
Centro Nacional de Investigaciones Metalurgicas	spectrometry standards

Reference Material Producer	List of Products
Chem Service USA	standards for GC, GC-MS, LC, TLC, NMR, IR, spectral analysis, microanalysis, environmenal analysis, pesticides, PCBs
Chemicals Inspection and Testing Institute, Japan	gas mixtures, cation/anion reference solutions, pH standards, serum, cholesterol
China Metalurgical Standardization Research Institute	metals, alloys, minerals
Chiron Laboratories, AS	petroleum standards, PAHs
Conoco Inc	AA, ICP, XRF and DCP standards
CTIF Centre Technique de la Fonderie	metals, alloys
Dow USA	magnesium spectrographic standard
Goodfellow Metals Ltd	metals, alloys, ceramics and other materials
Government Chemical Industrial Research Institute, Tokyo (MITI)	glass, rock, nuclear, volumetric, viscosity, metals, fuel, sea water
Haltermann	pure and technical hydrocarbons, ASTM reference fuels
Heraeus	organic, inorganic fine chemicals, pure elements, air products, stable isotopes
High Pressure Gas Safety Institute of Japan	high pressure gases
Imphy (UK) Ltd, High Fidelity Alloys	alloys
Institute du Verre, Paris	glass for chemical analysis and calibration of viscometers
Institute of Nuclear Chemistry & Technology	fly ash, tobacco leaves, apatite concentration
Instytut Przemyslu Organicznego, Poland	pesticide CRMs
IPT Instituto de Pesquisas Technologicas de Estado	metals, steels, alloys, refractories, ores, minerals, rocks
Iron & Steel Institute of Japan	iron, steel
J T Baker (UK)	spectrometric solutions
Korea Standards Research Institute	gases, steels and non-ferrous alloys, environmental
Laboratoires Rhone Alpes Mercure, France	high purity metals, electronic grade mercury, silver, gold, gallium
LRCCP	rubbers, plastics
MBH Reference Materials	metals, alloys, powders/chippings/solutions, rocks, soils, clay, ash
Micrographics	calibration standards for the petroleum, plastics, petrochemical and environmental industries
Ministry of International Trade and Industry, Japan	standard metal solutions
Nordisk Analys Service AB, Sweden	steel certified for oxygen, nitrogen, hydrogen, sulfur and carbon content
Philips Analytical	aluminium alloy CRMs for emission spectrometry
Physikalisch-Technische Bundesanstalt (PTB)	RMs for calibration services for magnetic media, radioactivity standards
Polymer Laboratories	polymer standards and columns

Reference Material Producer	*List of Products*
PRE	refractory materials, clay, shale
Pressure Chemical Co	polymer standards and instrument calibrants
Rapra Technology Ltd	rubbers and plastics
Referensmaterial AB	alloys, steels, ore, slag, gases in metals, standard aqueous solutions for AA and FEs
Riedel-deHaen	high purity, pesticides, solvents, polluting chemical standards, reference substances for gas chromatography
RSA Le Rubis SA	sapphire, ruby, spinels
Rutherford Research Products Co., N.J., USA	standard dielectric specimens and fluids
Société Nouvelle du Littoral	reference cement and alumina
Society of Glass Technology	sand, fluoride and opal glass, various glasses
Speciality Gases Ltd	analysed gases/gas mixtures
Spex	standards and compounds for inorganic spectroscopy
Standards Association of Australia	float glass, electrode carbon, iron ore, zircon sand, coal
Supelco	chromatography standards
Swedish Institute for Metals Research	steels, ferro alloys, iron ores, slags, fluorspars
USSR Association of Reference Material Producers	metallurgy, soils, rocks, oil, petroleum, geochemical, biological, physical properties
USSR Certified Reference Materials	metals and alloys
VASKUT, Hungary	steels, alloys *etc.*
Zaclad Doswiadczalny (Chemipan)	high purity organic compounds for chromatography

4. Biomedical

Amersham	bacteriology, immunoserology, radioimmunology, clinical chemistry, hemostasis
Asean Reference Substances	pharmaceuticals
ATCC (American Type Culture Collection)	cultures, bacteriological
bioMerieux (UK) Ltd	drugs and pharmaceuticals
College of American Pathologists	biological and pharmaceutical products
DHSS British Pharmacopoeia Commission Laboratory	serum cholesterol, HIV/hepatitis b, cocaine/marijuana standard solutions, proteins
National Institute for Biological Standards and Control, NIBSC, UK	biological standards and reference materials
Pharmacia	pharmaceuticals
SVM	biological, clinical
ThetaKits	pharmaceutical/drug standards
United States Phamacopeial Convention Inc	serum, vaccines
Wellcome Foundation	biological products

Reference Material Producer	List of Products
World Health Organisation	biological and pharmaceutical products

5. Physical Properties *etc.*

American Petroleum Institute	refractive index, density, calorimetric heat of combustion, hydrocarbons, organic sulfur and nitrogen compounds
Bureau National de Metrologie	metals and alloys, semiconductors, elements, minerals, gases, inorganic compounds, uranium compounds, density, viscosity, molecular mass pH, temp, combustion
Cannon Instrument Company, USA	viscosity standards
CSIRO – Division of Applied Physics, Australia	viscosity, certified oils
Czecho-Slovak Institute of Metrology	pH, titrimetry, conductivity, ionometry, organics, pesticides, carbon steel, metals, slags
Duke Scientific Corporation	particle size
Institute of Gas Technology (IGT)	high methane gases with certified calorific value, specific gravity and composition
Kodak Laboratory and Research Products	monodisperse microsphere standards, liquid density standards, neutron activation standards, AA standards, gelatin matrix trace element standards, white reflectance standards
NPL, UK	melting point, enthalpy of fusion, gas mixtures, ferrous alloys, quenching fluid, colorimetric/spectrophotometric, radionuclides, temperature standards, thermal conductivity, triple point cells
Optical Activity Ltd	polarimeter and temperature controllable quartz rotation standards
Radiometer, Copenhagen	pH, ion and conductivity measurements
Skogsindustrins Tekniska Forskningsinstitut (Swedish Pulp and Paper Research Institute)	RMs for calibration of instruments used for the properties of paper
Poulten, Selfe & Lee Ltd	viscosity standards
Cargille Laboratories USA	refractive index standards, sink-float standards, (liquid density monitoring), viscosity

6. Food & Agriculture

American Oil Chemists Society	fats, oils, related products
Larodan Fine Chemicals AB, Sweden	fatty acids, free fatty acids, n-hydrocarbons, lipids

Annex B   Statistical Tables

Table A1   *Critical Values for the Student t test*

υ 1T υ 2T	60% 20%	75% 50%	80% 60%	85% 70%	90% 80%	95% 90%	97.5% 95%	99% 98%	99.5% 99%	99.9% 99.8%	99.95% 99.9%
1	0.3249	1.0000	1.3764	1.963	3.078	6.314	12.706	31.821	63.657	318.3	636.62
2	0.2887	0.8165	1.0607	1.386	1.886	2.920	4.303	6.965	9.925	22.33	31.596
3	0.2767	0.7649	0.9785	1.250	1.638	2.353	3.182	4.541	5.841	10.21	12.941
4	0.2707	0.7407	0.9410	1.190	1.533	2.132	2.776	3.747	4.604	7.173	8.610
5	0.2672	0.7267	0.9195	1.156	1.476	2.015	2.571	3.365	4.032	5.893	6.869
6	0.2648	0.7176	0.9057	1.134	1.440	1.943	2.447	3.143	3.707	5.208	5.959
7	0.2632	0.7111	0.8960	1.119	1.415	1.895	2.365	2.998	3.499	4.785	5.408
8	0.2619	0.7064	0.8889	1.108	1.397	1.860	2.306	2.896	3.355	4.501	5.041
9	0.2610	0.7027	0.8834	1.100	1.383	1.833	2.262	2.821	3.250	4.297	4.781
10	0.2602	0.6998	0.8791	1.093	1.372	1.812	2.228	2.764	3.169	4.144	4.587
11	0.2596	0.6974	0.8755	1.088	1.363	1.796	2.201	2.718	3.106	4.025	4.437
12	0.2590	0.6955	0.8726	1.083	1.356	1.782	2.179	2.681	3.055	3.930	4.318
13	0.2586	0.6938	0.8702	1.079	1.350	1.771	2.160	2.650	3.012	3.852	4.221
14	0.2582	0.6924	0.8681	1.076	1.345	1.761	2.145	2.624	2.977	3.787	4.140
15	0.2579	0.6912	0.8662	1.074	1.341	1.753	2.131	2.602	2.947	3.733	4.073
16	0.2576	0.6901	0.8647	1.071	1.337	1.746	2.120	2.583	2.921	3.686	4.015
17	0.2573	0.6892	0.8633	1.069	1.333	1.740	2.110	2.567	2.898	3.646	3.965
18	0.2571	0.6884	0.8620	1.067	1.330	1.734	2.101	2.552	2.878	3.610	3.922
19	0.2569	0.6876	0.8610	1.066	1.328	1.729	2.093	2.539	2.861	3.579	3.883
20	0.2567	0.6870	0.8600	1.064	1.325	1.725	2.086	2.528	2.845	3.552	3.850
21	0.2566	0.6864	0.8591	1.063	1.323	1.721	2.080	2.518	2.831	3.527	3.819
22	0.2564	0.6858	0.8583	1.061	1.321	1.717	2.074	2.508	2.819	3.505	3.792
23	0.2563	0.6853	0.8575	1.060	1.319	1.714	2.069	2.500	2.807	3.485	3.768
24	0.2562	0.6848	0.8569	1.059	1.318	1.711	2.064	2.492	2.797	3.467	3.745
25	0.2561	0.6844	0.8562	1.058	1.316	1.708	2.060	2.485	2.787	3.450	3.725
26	0.2560	0.6840	0.8557	1.058	1.315	1.706	2.056	2.479	2.779	3.435	3.707
27	0.2559	0.6837	0.8551	1.057	1.314	1.703	2.052	2.473	2.771	3.421	3.690
28	0.2558	0.6834	0.8546	1.056	1.313	1.701	2.048	2.467	2.763	3.408	3.674
29	0.2557	0.6830	0.8542	1.055	1.311	1.699	2.045	2.462	2.756	3.396	3.659
30	0.2556	0.6828	0.8538	1.055	1.310	1.697	2.042	2.457	2.750	3.385	3.646
31	0.2556	0.6825	0.8534	1.054	1.309	1.695	2.039	2.453	2.744	3.375	3.634
32	0.2555	0.6822	0.8530	1.054	1.309	1.694	2.037	2.449	2.738	3.365	3.622
33	0.2554	0.6820	0.8527	1.053	1.308	1.692	2.034	2.445	2.733	3.356	3.611

Where **1T** represent a one-tailed test and **2T** represents a two-tailed test.

υ 1T υ 2T	60% 20%	75% 50%	80% 60%	85% 70%	90% 80%	95% 90%	97.5% 95%	99% 98%	99.5% 99%	99.9% 99.8%	99.95% 99.9%
34	0.2553	0.6818	0.8523	1.052	1.307	1.691	2.032	2.441	2.728	3.348	3.601
35	0.2553	0.6816	0.8520	1.052	1.307	1.690	2.030	2.437	2.723	3.340	3.591
36	0.2552	0.6814	0.8517	1.052	1.306	1.688	2.028	2.434	2.719	3.333	3.582
37	0.2551	0.6812	0.8515	1.051	1.305	1.687	2.026	2.431	2.715	3.326	3.574
38	0.2551	0.6810	0.8512	1.051	1.304	1.686	2.024	2.429	2.712	3.319	3.566
39	0.2550	0.6809	0.8509	1.051	1.304	1.685	2.022	2.426	2.708	3.313	3.558
40	0.2550	0.6807	0.8507	1.050	1.303	1.684	2.021	2.423	2.704	3.307	3.551
41	0.2550	0.6805	0.8505	1.050	1.303	1.683	2.019	2.421	2.701	3.301	3.544
42	0.2549	0.6803	0.8503	1.050	1.302	1.683	2.017	2.419	2.698	3.296	3.537
43	0.2549	0.6802	0.8501	1.049	1.302	1.682	2.016	2.417	2.695	3.291	3.531
44	0.2549	0.6800	0.8499	1.049	1.301	1.681	2.015	2.415	2.692	3.286	3.525
45	0.2548	0.6799	0.8497	1.049	1.301	1.680	2.014	2.413	2.689	3.281	3.519
46	0.2548	0.6798	0.8496	1.048	1.300	1.679	2.013	2.411	2.686	3.277	3.514
47	0.2548	0.6797	0.8494	1.048	1.300	1.678	2.012	2.409	2.684	3.273	3.509
48	0.2547	0.6796	0.8492	1.048	1.299	1.677	2.011	2.407	2.682	3.269	3.504
49	0.2547	0.6795	0.8490	1.048	1.299	1.676	2.010	2.405	2.680	3.265	3.500
50	0.2547	0.6794	0.8489	1.047	1.299	1.676	2.009	2.403	2.678	3.261	3.496
51	0.2547	0.6794	0.8487	1.047	1.298	1.675	2.008	2.402	2.676	3.257	3.492
52	0.2546	0.6793	0.8486	1.047	1.298	1.675	2.007	2.400	2.674	3.254	3.488
53	0.2546	0.6792	0.8485	1.047	1.298	1.674	2.006	2.398	2.672	3.251	3.484
54	0.2546	0.6791	0.8483	1.046	1.297	1.674	2.005	2.396	2.670	3.248	3.480
55	0.2546	0.6790	0.8482	1.046	1.297	1.673	2.004	2.395	2.668	3.245	3.476
56	0.2546	0.6789	0.8481	1.046	1.297	1.673	2.003	2.394	2.666	3.242	3.472
57	0.2545	0.6788	0.8480	1.046	1.296	1.672	2.002	2.393	2.664	3.239	3.469
58	0.2545	0.6787	0.8479	1.045	1.296	1.672	2.001	2.392	2.662	3.236	3.466
59	0.2545	0.6786	0.8478	1.045	1.296	1.671	2.000	2.391	2.661	3.234	3.463
60	0.2545	0.6786	0.8477	1.045	1.296	1.671	2.000	2.390	2.660	3.232	3.460
61	0.2545	0.6785	0.8476	1.045	1.296	1.671	1.999	2.389	2.659	3.230	3.457
62	0.2545	0.6785	0.8475	1.045	1.295	1.670	1.999	2.388	2.658	3.228	3.454
63	0.2544	0.6784	0.8474	1.045	1.295	1.670	1.998	2.387	2.657	3.226	3.451
64	0.2544	0.6784	0.8473	1.045	1.295	1.670	1.998	2.386	2.656	3.224	3.448
65	0.2544	0.6783	0.8472	1.045	1.295	1.669	1.997	2.385	2.655	3.222	3.445
66	0.2544	0.6783	0.8471	1.045	1.295	1.669	1.997	2.384	2.654	3.220	3.443
67	0.2544	0.6782	0.8470	1.045	1.295	1.669	1.996	2.383	2.653	3.218	3.441
68	0.2544	0.6782	0.8470	1.044	1.294	1.668	1.996	2.382	2.652	3.216	3.439
69	0.2544	0.6781	0.8469	1.044	1.294	1.668	1.995	2.382	2.651	3.214	3.437
70	0.2543	0.6781	0.8469	1.044	1.294	1.668	1.995	2.381	2.650	3.212	3.435
71	0.2543	0.6780	0.8468	1.044	1.294	1.668	1.994	2.381	2.649	3.210	3.433
72	0.2543	0.6780	0.8468	1.044	1.294	1.667	1.994	2.380	2.648	3.208	3.431
73	0.2543	0.6779	0.8467	1.044	1.294	1.667	1.993	2.380	2.647	3.207	3.429
74	0.2543	0.6779	0.8467	1.044	1.294	1.667	1.993	2.379	2.646	3.206	3.427
75	0.2543	0.6778	0.8466	1.044	1.293	1.667	1.992	2.379	2.645	3.205	3.425
76	0.2543	0.6778	0.8466	1.044	1.293	1.666	1.992	2.378	2.644	3.204	3.423
77	0.2543	0.6777	0.8465	1.044	1.293	1.666	1.992	2.378	2.643	3.203	3.421
78	0.2542	0.6777	0.8465	1.044	1.293	1.666	1.991	2.377	2.642	3.202	3.419
79	0.2542	0.6777	0.8464	1.043	1.293	1.666	1.991	2.377	2.641	3.201	3.417
80	0.2542	0.6776	0.8464	1.043	1.293	1.665	1.991	2.376	2.640	3.200	3.415
81	0.2542	0.6776	0.8463	1.043	1.293	1.665	1.990	2.376	2.639	3.199	3.413
82	0.2542	0.6776	0.8463	1.043	1.293	1.665	1.990	2.375	2.638	3.198	3.411
83	0.2542	0.6775	0.8462	1.043	1.292	1.665	1.990	2.375	2.637	3.197	3.410
84	0.2542	0.6775	0.8462	1.043	1.292	1.664	1.989	2.374	2.636	3.196	3.409

υ 1T υ 2T	60% 20%	75% 50%	80% 60%	85% 70%	90% 80%	95% 90%	97.5% 95%	99% 98%	99.5% 99%	99.9% 99.8%	99.95% 99.9%
85	0.2542	0.6775	0.8461	1.043	1.292	1.664	1.989	2.374	2.635	3.195	3.408
86	0.2542	0.6774	0.8461	1.043	1.292	1.664	1.989	2.373	2.634	3.194	3.407
87	0.2541	0.6774	0.8460	1.043	1.292	1.664	1.988	2.373	2.633	3.193	3.406
88	0.2541	0.6774	0.8460	1.043	1.292	1.664	1.988	2.372	2.633	3.192	3.405
89	0.2541	0.6773	0.8459	1.043	1.292	1.663	1.988	2.372	2.632	3.191	3.404
90	0.2541	0.6773	0.8459	1.043	1.292	1.663	1.987	2.371	2.632	3.190	3.403
91	0.2541	0.6773	0.8458	1.043	1.291	1.663	1.987	2.371	2.631	3.189	3.402
92	0.2541	0.6772	0.8458	1.042	1.291	1.663	1.987	2.370	2.631	3.188	3.401
93	0.2541	0.6772	0.8457	1.042	1.291	1.663	1.986	2.370	2.630	3.187	3.400
94	0.2541	0.6772	0.8457	1.042	1.291	1.662	1.986	2.369	2.630	3.186	3.399
95	0.2541	0.6771	0.8456	1.042	1.291	1.662	1.986	2.369	2.629	3.185	3.398
96	0.2541	0.6771	0.8456	1.042	1.291	1.662	1.986	2.368	2.629	3.184	3.397
97	0.2540	0.6771	0.8455	1.042	1.291	1.662	1.985	2.368	2.628	3.183	3.396
98	0.2540	0.6770	0.8455	1.042	1.291	1.662	1.985	2.367	2.628	3.182	3.395
99	0.2540	0.6770	0.8454	1.042	1.291	1.661	1.985	2.367	2.627	3.181	3.394
100	0.2540	0.6770	0.8454	1.042	1.290	1.661	1.985	2.366	2.627	3.180	3.393
∞	0.2533	0.6745	0.8416	1.036	1.282	1.645	1.960	2.326	2.576	3.090	3.291

Table A2 *97.5% Critical Values for the F-test (one-tailed, equivalent to 95% two-tailed)*

v_1 v_2	1	2	3	4	5	6	7	8	10	12	24	∞
1	647.8	799.5	864.2	899.6	921.8	937.1	948.2	956.7	968.6	976.7	997.2	1018.0
2	38.51	39.00	39.17	39.25	39.30	39.33	39.36	39.37	39.40	39.41	39.46	39.50
3	17.44	16.04	15.44	15.10	14.88	14.71	14.62	14.54	14.42	14.34	14.12	13.90
4	12.22	10.65	9.979	9.605	9.364	9.197	9.074	8.980	8.844	8.751	8.511	8.257
5	10.01	8.434	7.764	7.388	7.146	6.978	6.853	6.757	6.619	6.525	6.278	6.015
6	8.813	7.260	6.599	6.227	5.988	5.820	5.695	5.600	5.461	5.366	5.117	4.849
7	8.073	6.542	5.890	5.523	5.285	5.119	4.995	4.899	4.761	4.666	4.415	4.142
8	7.571	6.059	5.416	5.053	4.817	4.652	4.529	4.433	4.295	4.200	3.947	3.670
9	7.209	5.715	5.078	4.718	4.484	4.320	4.197	4.102	3.964	3.868	3.614	3.333
10	6.937	5.456	4.826	4.468	4.236	4.072	3.950	3.855	3.717	3.621	3.365	3.080
11	6.724	5.256	4.630	4.275	4.044	3.881	3.759	3.664	3.526	3.430	3.173	2.883
12	6.554	5.096	4.474	4.121	3.891	3.728	3.607	3.512	3.374	3.277	3.019	2.725
13	6.414	4.965	4.347	3.996	3.767	3.604	3.483	3.388	3.250	3.153	2.893	2.595
14	6.298	4.857	4.242	3.892	3.663	3.501	3.380	3.285	3.147	3.050	2.789	2.487
15	6.200	4.765	4.153	3.804	3.576	3.415	3.293	3.199	3.060	2.963	2.701	2.395
16	6.115	4.687	4.077	3.729	3.502	3.341	3.219	3.125	2.986	2.889	2.625	2.316
17	6.042	4.619	4.011	3.665	3.438	3.277	3.156	3.061	2.922	2.825	2.560	2.247
18	5.978	4.560	3.954	3.608	3.382	3.221	3.100	3.005	2.866	2.769	2.503	2.187
19	5.922	4.508	3.903	3.559	3.333	3.172	3.051	2.956	2.817	2.720	2.452	2.133
20	5.871	4.461	3.859	3.515	3.289	3.128	3.007	2.913	2.774	2.676	2.408	2.085
21	5.827	4.420	3.819	3.475	3.250	3.090	2.969	2.874	2.735	2.637	2.368	2.042
22	5.786	4.383	3.783	3.440	3.215	3.055	2.934	2.839	2.700	2.602	2.331	2.003
23	5.750	4.349	3.750	3.408	3.183	3.023	2.902	2.808	2.668	2.570	2.299	1.968
24	5.717	4.319	3.721	3.379	3.155	2.995	2.874	2.779	2.640	2.541	2.269	1.935
25	5.686	4.291	3.694	3.353	3.129	2.969	2.848	2.753	2.613	2.515	2.242	1.906
26	5.659	4.265	3.670	3.329	3.105	2.945	2.824	2.729	2.590	2.491	2.217	1.878
27	5.633	4.242	3.647	3.307	3.083	2.923	2.802	2.707	2.568	2.469	2.195	1.853
28	5.610	4.221	3.626	3.286	3.063	2.903	2.782	2.687	2.547	2.448	2.174	1.829
29	5.588	4.201	3.607	3.267	3.044	2.884	2.763	2.669	2.529	2.430	2.154	1.807
30	5.568	4.182	3.589	3.250	3.026	2.867	2.746	2.651	2.511	2.412	2.136	1.787
32	5.531	4.149	3.557	3.218	2.995	2.836	2.715	2.620	2.480	2.381	2.103	1.750
34	5.499	4.120	3.529	3.191	2.968	2.808	2.688	2.593	2.453	2.353	2.075	1.717
36	5.471	4.094	3.505	3.167	2.944	2.785	2.664	2.569	2.429	2.329	2.049	1.687
38	5.446	4.071	3.483	3.145	2.923	2.763	2.643	2.548	2.407	2.307	2.027	1.661
40	5.424	4.051	3.463	3.126	2.904	2.744	2.624	2.529	2.388	2.288	2.007	1.637
60	5.286	3.925	3.343	3.008	2.786	2.627	2.507	2.412	2.270	2.169	1.882	1.482
120	5.152	3.805	3.227	2.894	2.674	2.515	2.395	2.299	2.157	2.055	1.760	1.310
∞	5.024	3.689	3.116	2.786	2.567	2.408	2.288	2.192	2.048	1.945	1.640	1.000

Table A3 *95% Critical Values for the F-test (one-tailed)*

v_1 / v_2	1	2	3	4	5	6	7	8	10	12	24	∞
1	161.4	199.5	215.7	224.6	230.2	234.0	236.8	238.9	241.9	243.9	249.1	254.3
2	18.51	19.00	19.16	19.26	19.30	19.33	19.35	19.37	19.40	19.41	19.45	19.50
3	10.13	9.552	9.277	9.117	9.013	8.941	8.887	8.845	8.786	8.745	8.639	8.526
4	7.709	6.944	6.591	6.388	6.256	6.163	6.094	6.041	5.964	5.912	5.774	5.628
5	6.608	5.786	5.409	5.192	5.050	4.950	4.876	4.818	4.735	4.678	4.527	4.365
6	5.987	5.143	4.757	4.534	4.387	4.284	4.207	4.147	4.060	4.000	3.841	3.669
7	5.591	4.737	4.347	4.120	3.972	3.833	3.787	3.726	3.637	3.575	3.410	3.230
8	5.318	4.459	4.066	3.838	3.687	3.581	3.500	3.438	3.347	3.284	3.115	2.928
9	5.117	4.256	3.863	3.633	3.482	3.374	3.293	3.230	3.137	3.073	2.900	2.707
10	4.965	4.103	3.708	3.478	3.326	3.217	3.135	3.072	2.978	2.913	2.737	2.538
11	4.844	3.982	3.587	3.357	3.204	3.095	3.012	2.948	2.854	2.788	2.609	2.404
12	4.747	3.885	3.490	3.259	3.106	2.996	2.913	2.849	2.753	2.687	2.505	2.296
13	4.667	3.806	3.411	3.179	3.025	2.915	2.832	2.767	2.671	2.604	2.420	2.206
14	4.600	3.739	3.344	3.112	2.958	2.848	2.764	2.699	2.602	2.534	2.349	2.131
15	4.543	3.682	3.287	3.056	2.901	2.790	2.707	2.641	2.544	2.475	2.288	2.066
16	4.494	3.634	3.239	3.007	2.852	2.741	2.657	2.591	2.494	2.425	2.235	2.010
17	4.451	3.592	3.197	2.965	2.810	2.699	2.614	2.548	2.450	2.381	2.190	1.960
18	4.414	3.555	3.160	2.928	2.773	2.661	2.577	2.510	2.412	2.342	2.150	1.917
19	4.381	3.522	3.127	2.895	2.740	2.628	2.544	2.477	2.378	2.308	2.114	1.878
20	4.351	3.493	3.098	2.866	2.711	2.599	2.514	2.447	2.348	2.278	2.082	1.843
21	4.325	3.467	3.072	2.840	2.685	2.573	2.488	2.420	2.321	2.250	2.054	1.812
22	4.301	3.443	3.049	2.817	2.661	2.549	2.464	2.397	2.297	2.226	2.028	1.783
23	4.279	3.422	3.028	2.796	2.640	2.528	2.442	2.375	2.275	2.204	2.005	1.757
24	4.260	3.403	3.009	2.776	2.621	2.508	2.423	2.355	2.255	2.183	1.984	1.733
25	4.242	3.385	2.991	2.759	2.603	2.490	2.405	2.337	2.236	2.165	1.964	1.711
26	4.225	3.369	2.975	2.743	2.587	2.474	2.388	2.321	2.220	2.148	1.946	1.691
27	4.210	3.354	2.690	2.728	2.572	2.459	2.373	2.305	2.204	2.132	1.930	1.672
28	4.196	3.340	2.947	2.714	2.558	2.445	2.359	2.291	2.190	2.118	1.915	1.654
29	4.183	3.328	2.934	2.701	2.545	2.432	2.346	2.278	2.177	2.104	1.901	1.638
30	4.171	3.316	2.922	2.690	2.534	2.421	2.334	2.266	2.165	2.092	1.887	1.622
32	4.149	3.295	2.901	2.668	2.512	2.399	2.313	2.244	2.142	2.070	1.864	1.594
34	4.130	3.276	2.883	2.650	2.494	2.380	2.294	2.225	2.123	2.050	1.843	1.569
36	4.113	3.259	2.866	2.634	2.477	2.364	2.277	2.209	2.106	2.033	1.824	1.547
38	4.098	3.245	2.852	2.619	2.463	2.349	2.262	2.194	2.091	2.017	1.808	1.527
40	4.085	3.232	2.839	2.606	2.449	2.336	2.249	2.180	2.077	2.003	1.793	1.509
60	4.001	3.150	2.758	2.525	2.368	2.254	2.167	2.097	1.993	1.917	1.700	1.389
120	3.920	3.072	2.680	2.447	2.290	2.175	2.087	2.016	1.910	1.834	1.608	1.254
∞	3.841	2.996	2.605	2.372	2.214	2.099	2.010	1.938	1.831	1.752	1.517	1.000

Table A4   *Cumulative Distribution of Chi-squared*

df	95%	99%	df	95%	99%	df	95%	99%
1	3.842	6.635	51	68.683	77.408	101	125.492	137.023
2	5.992	9.211	52	69.846	78.638	102	126.608	138.187
3	7.815	11.346	53	71.008	79.866	103	127.723	139.350
4	9.488	13.278	54	72.168	81.092	104	128.838	140.513
5	11.071	15.088	55	73.326	82.316	105	129.953	141.675
6	12.593	16.814	56	74.484	83.538	106	131.067	142.836
7	14.068	18.478	57	75.639	84.758	107	132.180	143.996
8	15.509	20.093	58	76.794	85.976	108	133.293	145.156
9	16.921	21.669	59	77.947	87.192	109	134.406	146.314
10	18.309	23.213	60	79.099	88.406	110	135.517	147.472
11	19.677	24.729	61	80.249	89.619	111	136.629	148.630
12	21.028	26.221	62	81.398	90.830	112	137.740	149.786
13	22.365	27.693	63	82.546	92.039	113	138.850	150.942
14	23.688	29.146	64	83.693	93.246	114	139.960	152.098
15	24.999	30.583	65	84.839	94.452	115	141.069	153.252
16	26.299	32.006	66	85.984	95.656	116	142.178	154.406
17	27.591	33.415	67	87.127	96.859	117	143.287	155.559
18	28.873	34.812	68	88.270	98.060	118	144.395	156.712
19	30.147	36.198	69	89.411	99.259	119	145.502	157.864
20	31.415	37.574	70	90.552	100.458	120	146.609	159.015
21	32.675	38.940	71	91.691	101.655	121	147.716	160.166
22	33.929	40.298	72	92.829	102.850	122	148.822	161.316
23	35.177	41.647	73	93.967	104.044	123	149.928	162.466
24	36.420	42.989	74	95.103	105.237	124	151.033	163.614
25	37.658	44.324	75	96.239	106.428	125	152.138	164.763
26	38.891	45.652	76	97.374	107.619	126	153.243	165.910
27	40.119	46.973	77	98.508	108.808	127	154.347	167.058
28	41.343	48.289	78	99.640	109.995	128	155.450	168.204
29	42.564	49.599	79	100.773	111.182	129	156.554	169.350
30	43.780	50.904	80	101.904	112.367	130	157.657	170.495
31	44.993	52.203	81	103.034	113.552	131	158.759	171.640
32	46.202	53.498	82	104.164	114.735	132	159.861	172.785
33	47.408	54.789	83	105.293	115.917	133	160.963	173.928
34	48.610	56.074	84	106.421	117.098	134	162.064	175.072
35	49.810	57.356	85	107.548	118.277	135	163.165	176.214
36	51.007	58.634	86	108.675	119.456	136	164.266	177.356
37	52.201	59.907	87	109.800	120.634	137	165.366	178.498
38	53.393	61.177	88	110.926	121.811	138	166.466	179.639
39	54.582	62.444	89	112.050	122.986	139	167.565	180.780
40	55.768	63.707	90	113.174	124.161	140	168.665	181.920
41	56.953	64.967	91	114.297	125.335	141	169.763	183.060
42	58.135	66.224	92	115.419	126.508	142	170.862	184.199
43	59.314	67.477	93	116.541	127.680	143	171.960	185.337
44	60.492	68.728	94	117.662	128.851	144	173.058	186.476
45	61.668	69.976	95	118.782	130.021	145	174.155	187.613
46	62.841	71.221	96	119.902	131.190	146	175.252	188.751
47	64.013	72.463	97	121.021	132.358	147	176.349	189.887
48	65.183	73.703	98	122.140	133.526	148	177.445	191.024
49	66.351	74.940	99	123.258	134.692	149	178.542	192.160
50	67.518	76.175	100	124.375	135.858	150	179.637	193.295

Table A5 *Grubbs' Critical Value Table*

n	95% confidence level			99% confidence level		
	G_1	G_2	G_3	G_1	G_2	G_3
3	1.153	2.00	—	1.155	2.00	—
4	1.463	2.43	0.9992	1.492	2.44	1.0000
5	1.672	2.75	0.9817	1.749	2.80	0.9965
6	1.822	3.01	0.9436	1.944	3.10	0.9814
7	1.938	3.22	0.8980	2.097	3.34	0.9560
8	2.032	3.40	0.8522	2.221	3.54	0.9250
9	2.110	3.55	0.8091	2.323	3.72	0.8918
10	2.176	3.68	0.7695	2.410	3.88	0.8586
12	2.285	3.91	0.7004	2.550	4.13	0.7957
13	2.331	4.00	0.6705	2.607	4.24	0.7667
15	2.409	4.17	0.6182	2.705	4.43	0.7141
20	2.557	4.49	0.5196	2.884	4.79	0.6091
25	2.663	4.73	0.4505	3.009	5.03	0.5320
30	2.745	4.89	0.3992	3.103	5.19	0.4732
35	2.811	5.026	0.3595	3.178	5.326	0.4270
40	2.866	5.150	0.3276	3.240	5.450	0.3896
50	2.956	5.350	0.2797	3.336	5.650	0.3328
60	3.025	5.500	0.2450	3.411	5.800	0.2914
70	3.082	5.638	0.2187	3.471	5.938	0.2599
80	3.130	5.730	0.1979	3.521	6.030	0.2350
90	3.171	5.820	0.1810	3.563	6.120	0.2147
100	3.207	5.900	0.1671	3.600	6.200	0.1980
110	3.239	5.968	0.1553	3.632	6.268	0.1838
120	3.267	6.030	0.1452	3.662	6.330	0.1716
130	3.294	6.086	0.1364	3.688	6.386	0.1611
140	3.318	6.137	0.1288	3.712	6.437	0.1519

Subject Index